D0726315

RETAIL GEOGRAPHY AND INTELLIGENT NETWORK PLANNING

RETAIL GEOGRAPHY AND INTELLIGENT NETWORK PLANNING

MARK BIRKIN
University of Leeds, UK

GRAHAM CLARKE
University of Leeds, UK

and

MARTIN CLARKE
University of Leeds, UK and GMAP Ltd, UK

JOHN WILEY & SONS, LTD

Other Wiley Editorial Offices

John Wiley & Sons, Inc., 605 Third Avenue,
New York, NY 10158-0012, USA

Jossey-Bass, 989 Market Street, San Francisco,
CA 94103-1741, USA

Wiley-VCH Verlag GmbH,
Pappelallee 3, D-69469 Weinheim, Germany

John Wiley & Sons Australia, Ltd.,
33 Park Road, Milton, Queensland 4064, Australia

John Wiley & Sons (Asia) Pte Ltd., 2 Clementi Loop #02-01,
Jin Xing Distripark, Singapore 129809

John Wiley & Sons Canada Ltd., 22 Worcester Road,
Etobicoke, Ontario M9W 1L1

British Library Cataloguing in Publication Data

A catalogue record for this book is available from the British Library

ISBN 0 471 49761 4 (HB)   0 471 49803 3 (PB)

Typeset in 8.5/10pt Lucida Sans by Laserwords Private Limited, Chennai, India
Printed and bound in Great Britain by Biddles Ltd, Guildford and King's Lynn
This book is printed on acid-free paper responsibly manufactured from sustainable forestry in which at least two trees are planted for each one used for paper production.

Dedicated to

**Sir Alan Wilson,
friend and mentor**

CONTENTS

ACKNOWLEDGEMENTS

The authors are extremely grateful to David Appleyard and Adam Davenport for drawing many of the figures and tables. Lyn Roberts of Wiley also deserves our thanks for being so encouraging and yet patient as we missed many a deadline! Finally, our thanks go to all the employees of GMAP who have, at various times, contributed to the case studies presented here.

INTRODUCTION

1

The books, *Retail Location and Retail Planning* (Guy 1980) and *Retail Geography*, (Dawson 1981), were landmark publications in the study of spatial variations in demand and supply of retail activity. The importance of retail location was firmly established in the literature, and from then on, a large number of urban and economic geographers began to take retailing seriously. Other important and explicitly geographical retail texts followed in the 1980s and early 1990s (Davies 1984, Davies and Rogers 1984, Jones and Simmons 1987, 1990, Ghosh and MacLafferty 1987). Since then, there has been very little except for Guy (1994) and Wrigley and Lowe (1996). Twenty years on from those first texts, we feel it is time for an updated book on retail geography and strategic planning. The motivation for writing this book has come from three main directions. The first is in response to the so-called *new retail geography*, which has become popular in the last five years or so. Within this geography there seems little place for traditional concerns with retail location, especially when it is addressed through geographical information systems (GIS) and spatial modelling. The second is the attack on store location research from those who believe that few retailers will be opening new outlets in already crowded or saturated markets, especially since e-commerce has begun to make serious inroads into the sales of traditional outlets. The third, which is linked strongly to the second factor, is the view derived from our experience over the last ten years of the continued importance of location to all retail organizations. This has been built up over many years through our consultancy work, using GIS and spatial models (both in the School of Geography and in the private company GMAP Ltd). In the late 1990s, when it peaked, this work was worth around seven million pounds per annum and employed over one hundred geography graduates. This, to us at least, is evidence that location issues in retailing have never faded away, and indeed, we might argue, have never been as important as they are today. We shall address each of these issues in more detail below.

The 'new' retail geography is a term first applied by Wrigley and Lowe (1996). In a review of recent trends in retail geography, Crewe (2000) describes this as a 'reconstructed' retail geography, giving an account of the transformation from the old 'boring' geography, which 'misrepresented both the wider structure of the commodity channel and the status of consumption in shaping retail change' (cf. D. Clarke 1996). There is no doubt that this new retail geography is theoretically well informed and is an extremely important development. Indeed, we draw upon a lot of this literature in the review sections of the book. A summary of all recent work will appear in another very useful and complementary reader to this book by Wrigley and Lowe (2002). That said, there is clearly no place in this new retail geography agenda for traditional concerns with store location research, GIS and models. This is a great shame! The kitbag of techniques for handling store location issues has developed rapidly, and there are as many theoretical developments in this area as there are in cultural and economic retail geography (e.g. Birkin *et al.* 1996, Fotheringham *et al.* 2000, Longley *et al.* 2001, Clarke and Madden 2001). The topic of store location research has certainly moved on from the days of simply assessing the impacts of new store openings. Models are commonly used now to address one or more of the following strategic issues:

- evaluating existing branch performance (comparing expected sales in a locality with actual ones)
- impacts of new store openings (revenue predictions and impact on retailer and competitors)
- impacts of store closures
- impact of relocations
- assessing the returns from an increase/decrease in store size
- finding the optimal location for a new outlet
- comparing the present configuration of stores with an 'optimal' distribution
- presenting the best set of locations to target an overseas market
- launching a new product in locations most likely to maximize sales
- finding the best geographical fit for possible merger/acquisition candidates
- optimizing the branch networks of two or more organizations following merger/ acquisition
- defining store territories optimally (especially if required by law)
- assessing the likelihood of a merger or acquisition being referred to the authorities on the grounds of monopoly power
- maximizing the returns from different distribution channels in different locations
- impacts of changing the retail brand or fascia.

Examples of many of these will be presented in this book. Our argument is that what is increasingly required is greater intelligence on how to solve these problems. In an earlier work (Birkin *et al.* 1996), we argued for GIS to be more closely linked to spatial models in order to provide that intelligence. This volume continues from that, but concentrates on retail examples in order to more fully flesh out the concept of retail intelligence applied to network planning.

The argument against GIS and models is seemingly strengthened by recent events in retailing, which suggest that the traditional store is, or will be, less important in the retail scene of the future. To put it another way, as domestic markets are perceived to be getting more saturated, and because retailers are opening fewer new stores per year, the need for sophisticated store location research is getting less obvious (see Clarkson *et al.* 1996 for a good example of this viewpoint). In addition, according to many commentators, e-commerce will end the need for opening new outlets and will take a significant amount of trade away from the brick and mortar stores. Indeed, there is the belief that e-commerce will ultimately mean the death of the importance of distance and geography (cf. Cairncross 1997). Our response to this line of argument is that we believe that the techniques of spatial modelling are even more important in the present and in the future, not less. First, we have no doubt that the physical store is still going to be the main retail location of the future, despite what even the most optimistic e-commerce analysts believe, and it is evident from the press and company web pages that many companies are committed to yet more store openings, even in the crowded UK market (see Chapter 3 for more examples of planned growth). Second, retailers will have to face a variety of distribution channels in the future. They will need to find some way of maximizing sales from traditional stores, e-commerce, telephone sales and even perhaps automated branches. This means that retail growth strategies will be more complex. Is such complexity impossible to deal with and to plan effectively for? Far from throwing location techniques away, we argue that what retailers will require are more sophisticated and intelligent spatial analysis techniques capable of dealing with multichannel networks. In Chapters 6 to 10 we hope to demonstrate that geographers are up to the task.

The overriding motivation for this book is to show the usefulness of our techniques to the kinds of questions raised in the preceding text. In that sense, this is not a purely theoretical book that concludes that these techniques could be used in planning. The examples in the book are based on fifteen years of retail consultancy by each of the authors. During these years we have built up expertise in a wide range of modelling techniques relevant to the retail industry. It is reassuring to see that there is also a growing trend elsewhere in academia to directly work with retail clients (see Jones and Hernandez 2002 for many US examples, Scholten and Meijer 2002 for Dutch examples, and more generally see the papers in Clarke and Madden 2001, Clarke and Stillwell 2002). Yet, despite this growing link between academia and the real world, it is interesting to reflect on why such methods are not in widespread use. Despite the great advances in geographical data handling, it seems that many retailers still rely on gut feeling and good–fashioned retail nose (Hernandez 1998, Hernandez and Bennison 2000). Why is it so? Cost is likely to be a factor, but it cannot be the complete answer because the price of even a national GIS and modelling system would be a fraction of the annual marketing budgets of most retail organizations. A more plausible explanation is provided by Clarke I et al. (2000). They suggest that these more complex, sophisticated techniques either ignore or underplay the retailer's intuitive judgment in the location decision-making process. To this can also be added the observation that very often analysts do not want to support methodologies that might cast their previous decisions in poorer light. The response of Clarke I et al. (2000) is to try and capture the retail decision process more effectively and then to provide a framework that maps the 'cognitive and intuitive constructs underlying the schemas of retail executives' (p. 266). From this, a set of factors key to the site location decision should appear, which can then be used as an alternative to 'normative procedures'. Fair enough, but this framework is unlikely by itself to produce better or more informed decisions because the subjective factors that retail executives use still lie at the heart of these frameworks! Our response is to try and demonstrate through the examples in the second half of the book that the benefits of using more sophisticated spatial analysis tools outweigh the costs many times.

The rest of the book is set out as follows. In Chapter 2, we attempt a very broad-brush review of key retail trends. This has to be partial and selective, given space constraints, but we believe it is important as a backcloth to many of the examples given in Chapters 6 to 10. The review focuses on issues that are most likely to impact directly on retail location and store location research. In Chapter 3, we review what is happening to key retail destinations, and look at the emergence of new distribution channels, which potentially may threaten traditional outlets. Here, we have tried to reflect on past developments, but at the same time, we look ahead to see how important these destinations/new channels might be in the future. The focus in Chapter 4 switches to retail growth strategies and reviews growth from new store openings to growth through mergers, acquisitions, franchising and strategic alliances. This represents a very obvious geographical problem: where should we be and how shall we get there? In Chapter 5, given the rapidly growing literature, we take a deeper look at e-commerce and its growth. The argument we wish to make here is that geography will still be crucial to the success of e-commerce. This is because access to the technology of e-commerce is not uniform (and probably unlikely to ever be) and because retailers still have to get the goods to the consumer in a cost-effective manner.

In the second half of the book, we switch to spatial analysis methods and their usefulness in addressing retail growth strategies and the types of question listed in the preceding text. In Chapter 6, we begin with simple models of territory planning. This is important for a number of reasons. First, many organizations assign sales territories to their stores for operational management purposes. Second, in some cases, there is a legal requirement for stores to have an assigned territory that does not overlap with another sales area. Third, many retail organizations assign territories to warehouses for physical distribution purposes and/or area management/salesperson teams. This chapter introduces a number of methods for assigning territories, including GIS. This is followed in Chapter 7 by a broad review of store location methodologies, including deductive methods (which derive models from apriori theory of store performance) and inductive methods (which build theories of store performance based on the data). This chapter thus builds up a kitbag of methodologies in common practice, from gut feeling to more sophisticated spatial interaction approaches. The spatial interaction approach is described in more detail in Chapter 8. This has been our bread and butter methodology and we describe a number of case studies. We also argue that models often need to be highly disaggregated in order to replicate real-world customer flows. Chapter 9 looks at the issue of optimization and its usefulness in retail planning. Here we use two main examples: an optimization model that allows an organization to either add new stores in a sequential (but optimal) fashion or allows an entire network to be located optimally, given the usual variations in consumer demand and levels of competition. Direct marketing and geodemographics is the subject matter of Chapter 10. This is an increasingly important one because as the amount and quality of data increases, firms can now use the most up-to-date technology to profile their customers and then search out similar customers in other geographical locations. The benefits of analytical and model-based approaches to retail planning are summarized in Chapter 11. Some concluding comments are offered in Chapter 12. Inevitably, the book is biased to UK examples, but where possible we have added European and North American case studies.

The book is written for two types of readers. It might be optimistic, but we hope that people in retail location teams and consultancies will find much to interest them in terms of solving applied problems. Second, we hope geography and marketing students will find the material useful for their courses. However, more importantly, we wish to impress on these students that geography is a very applied discipline, and that the skills and techniques they learn in the classrooms can directly lead to jobs in retailing and marketing. Many of our former students are now in the retail business (either directly within companies or as part of consultancy teams). Indeed, some have even formed their own retail consultancy organizations. Read on and earn your fortune!

TRENDS IN RETAILING

2.1 Introduction

The aim of this chapter is to review broad trends both in consumer demand for goods and services and in major issues relating to the retail industry itself. In both cases, discussion will focus on those issues with special interest to the *geography* of retail change. It is always difficult to judge whether key trends are entirely demand-led or supply-led. The corner shop to supermarket transition in most western countries is a case in point. On the one hand, consumer pressures for cheaper prices and greater consumer mobility undoubtedly created conditions that allowed supermarkets to grow. On the other hand, their development cannot be understood without recourse to the rapid growth in retail buying power, resulting from the most innovative retailers gaining greater scale economies (which in turn allowed discount prices to be offered to the consumer). Thirdly, the transition required a pliant planning regime, both nationally and locally. Hence, retail change can involve a complex set of processes including demand, supply and third party activities.

In Section 2.2, we review broad trends in retail demand and retail consumption. An important aspect of contemporary retail geography is the drive to understand and set consumption in a more powerful theoretical framework (part of the so-called 'new' retail geography – see Chapter 1 for more discussion). A detailed exposition of all the new theoretical ideas on consumption takes us beyond the scope of this book. Crewe (2000), Lowe and Wrigley (2000) and Wrigley and Lowe (2002) provide a useful summary of this literature (see also Bell and Valentine 1997 and Corrigan 1997 for more details). The reader is recommended to supplement the brief review here with a detailed look at this growing literature.

In Section 2.3, we switch to key supply-side changes, focusing on concentration and globalization, cost cutting and saturation. Finally, in Section 2.4, we outline broad changes in planning legislation that are also key drivers of, or barriers to, retail development.

2.2 Demand-Side Changes

2.2.1 Introduction

Consumers are the lifeblood of the retail industry. Their preferences, their choices and their behaviour fundamentally influence the way in which the retail landscape evolves. It is therefore essential for all retailers to know how consumer characteristics are likely to change over the next few years and to develop strategies to exploit this change. Until fairly recently, retailers have been relatively insensitive to consumer needs. Shops and banks opened for restricted periods and general levels of customer service were poor. Ogbonna and Wilkinson (1990, p.10) quote a grocery store manager in the 1980s: 'Customer service was way down the list of priorities. All the company was really interested in was getting the shelves filled...' (also quoted in I.Clarke 2000). This has now all changed – the customer is king because customers have become more discerning, demanding greater levels of quality, service and convenience (Poyner 1987, Palmer and Beddall 1997). Levels of customer care have improved dramatically because retailers strive to generate high levels of customer satisfaction and loyalty. Those retailers who have not been able to keep up with changing consumer tastes and behaviours have struggled in the marketplace. Dawson (2000) suggests that the difficulties faced by Marks and Spencers in the 1990s are a result of the company losing touch with its customers.

The key question we now turn to is what have been the major dynamics of consumer behaviour? We shall explore a number of issues in the following sections.

2.2.2 Consumer mobility

Over the last 30 years, the activity patterns of consumers have changed quite dramatically. Individuals are spending more time travelling, both to work and to shop. Table 2.1 shows the increase in time spent travelling per week per person in the United Kingdom during the 1990s. As we can see, there has been a significant growth in the time spent travelling overall, and a significant increase in car mobility. Car ownership itself has grown considerably during this period in all European countries. Table 2.2 shows the current level of car ownership among EU member states. As one might expect, there is a striking correlation between car ownership rates and income (Luxembourg, Belgium, France, Germany and the United Kingdom). There is every reason to believe that car ownership rates will continue to increase as a result of increased prosperity. In the United Kingdom, for example, the Department of Transport estimates that the number of cars on the road will increase from 23 million in 1997 to 32 million by 2015. Despite the stated

Table 2.1 UK transport usage 1989 to 1998

	1989	1998
Air	90	125 (millions per annum)
Bus	44 300	43 200 (millions per km)
Train	33 406	35 400
Car	581 000	630 000

Source: Eurostat 2001.

Table 2.2 Car ownership per 1000 inhabitants 1998/1999

Luxembourg	572	UK	404
Italy	545	Netherlands	376
Germany	508	Denmark	343
France	456	Portugal	321
Belgium	440	Ireland	309
Spain	408		

Source: Eurostat

intention of all the governments of the EU to reduce car usage (with the aim of decreasing pollution and congestion), it is difficult to envisage anything other than a continued growth in car ownership in Europe. Table 2.2 shows countries such as Portugal are rapidly catching up with the car ownership rates of other European countries.

In some ways, changes in the retail system itself have been a driver behind the increased utilization of cars and have responded to seemingly increased preference by consumers for car-based journeys. The development of large out-of-town shopping malls in many European countries has generated long-distance shopping trips. For example, Figure 2.1 shows the catchment area for the Meadow Hall shopping centre close to Sheffield in the United Kingdom. As can be seen, the catchment area extends well over a one-hour journey time. Guy (1998b) notes that increasing mobility has not only allowed consumers greater choice of retail destination, but has also weakened the local monopolies of independent retailers, and allowed more aggressive newcomers to compete on the basis of convenient car access and parking.

The mobility explosion has also led to many interesting new developments in consumer behaviour. There has certainly been a growth in multipurpose trips. It is increasingly common for consumers to shop after work, helped by more liberal opening hours in most countries. It is estimated that the number of shopping trips in the United Kingdom made between 6 P.M. and 10 P.M. has risen from 6.4% of the total in 1985 to 15.0% today. However, we should bear in mind that not all social or cultural groups have enjoyed greater mobility. Van Kenhowe and De Wulf (2000) provide a useful typology of consumer groups based around the themes of money poor/rich and time poor/rich. The most mobile group is the money rich/time poor (largely 30 to 55 years of age, living with a partner and 2 or 3 older children, well educated and with a full-time job). They are the consumers most likely to plan multipurpose trips around hectic family schedules, and perhaps to enjoy e-commerce (see Chapter 5).

2.2.3 Expenditure & income

Table 2.3 shows the changing nature of consumer expenditure in one modern western economy (the United Kingdom). It is interesting to note how consumer expenditure patterns have shifted. The main trend shows a decline in the percentage share of expenditure on food, fuel and power and clothing and footwear. Households today are spending relatively more on household goods, travel and general retail and leisure services. The same patterns are true right across Europe (see the collection of consumer studies in Leefland and van

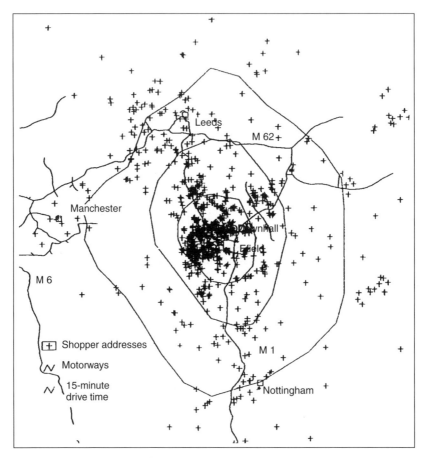

Figure 2.1 Using a geographic information system (GIS) to pinpoint customers. Source: OXIRM October 1990

Raaji, 1995). Again, they have interesting implications for retailers as consumers are spending more of their income on services and leisure.

Income is the crucial factor that drives expenditure. Generally, most countries have seen an overall increase in disposable income over the last few decades. For the United Kingdom, Green (1998) reports a substantial growth in real earnings (wages adjusted for price changes) since the 1970s. What is perhaps more interesting is that the higher income groups in many western countries have fared better over the last twenty years than lower income groups. This is a process referred to as social polarization and has applied to

Table 2.3 Changing nature of consumer markets

Expenditure patterns shifting...	1970	1988	1995	2000
Housing	12.6	17.5	16.4	16.0
Fuel & power	6.3	5.1	4.6	3.0
Food	25.7	18.7	17.8	17.0
Alcohol	4.5	4.5	4.3	4.0
Tobacco	4.8	2.2	2.0	2.0
Clothing & footwear	9.2	7.1	6.0	6.0
Household goods	6.5	7.3	8.0	9.0
Motoring & travel	13.6	14.8	15.1	18.0
Services	9.2	13.4	16.3	21.0
Leisure goods	n/a	4.7	4.9	5.0

Source: Family Expenditure Surveys (Various).

most western countries since the last war. The reason for this increasing polarization is the changing nature of the labour market. There has been a decline in manufacturing jobs in most western countries and an associated decline in skilled manual jobs. Instead, we have seen a growth in the service economy, which is divided into both highly skilled professional and managerial jobs and lower skilled lowly paid jobs (Pinch 1993). Thus, in effect, the traditional middle class has been squeezed in size, with an increase in both the upper and lower classes. Hamnett (1996) argues that the increase at the top end of the income bracket has been greater than the increase at the bottom end – thus, he prefers the term *professionalism* to polarization. Whichever term we use, it is clear that the geography of this widening income disparity is not even, with areas in the southeast of the United Kingdom greatly out-performing regions such as the southwest and the north (Green 1998). This has profound implications for all retailers, and especially those that target different income groups. Monitoring how much income and expenditure the target group has is an important marketing task. Understanding the geography of income gain and loss is also crucial. It is important that retailers have a sound methodology for estimating income and thus demand. Methods for estimating demand (using income and expenditure data) are given in Chapters 8 and 10.

2.2.4 Changing age profiles

Although it is well documented that the general trend across developed countries is a growth in the older population, some of the forecasted shifts in population are subtler. Table 2.4 shows how the age profile of a number of European countries is likely to change between now and the year 2010. What we shall witness is a substantial reduction in the 20- to 29-year-old population, a marked increase in the 50 to 64-year-old population, and in most countries a very sharp rise in the over 75s. Table 2.5 shows the key UK change figures, highlighting very high increases in the

Table 2.4 Change in key demographic cohorts 1995–2010

		20–29	50–64	75+
B	1995	1443	1643	620
	2010	1287	2035	889
	%	−11	23.8	43.4
DK	1995	781	857	364
	2010	610	1080	350
	%	−21.8	26.0	−3.8
D	1995	11964	15706	5118
	2010	9842	15886	6690
	%	−17.7	1.1	30.7
E	1995	6540	6241	2388
	2010	4699	7324	3585
	%	−28.1	17.4	50
F	1995	8578	8571	3531
	2010	7903	11845	5506
	%	−7.9	38.2	56.0
IRL	1995	538	462	173
	2010	605	664	197
	%	12.5	43.7	13.8
I	1995	9135	10304	3694
	2010	5984	11166	5745
	%	−34.5	8.4	55.5
NL	1995	2447	2335	848
	2010	2001	3344	1098
	%	018.2	43.3	27.9
A	1995	1285	1323	490
	2010	1029	1571	648
	%	−19.9	18.7	32.2
FIN	1995	661	805	292
	2010	646	1148	390
	%	−2.3	42.6	34.9
S	1995	1222	1420	723
	2010	1136	1735	802
	%	−7.0	22.2	10.9
UK	1995	8732	9079	4031
	2010	7764	11428	4482
	%	−11.1	25.9	11.2

Source: Evrostat.

population aged over 50 since the early 1990s. Across Europe, life expectancy has increased throughout the century. Perhaps, more important than the simple growth in life expectancy is the increase in expectancy at the age of 60. Not only are people living longer, but when they have reached a certain age, they can also expect to live

Table 2.5 Ageing UK population,
1994–2004

Age	Men	Women
0–14	−8	−7
15–49	−9	−11
50–74	+37	+30
75+	+88	+39

Source: Oxford Institute of Retail
Management.

considerably longer than they did 30 or 40 years ago. The reasons for increased life expectancy are both increased prosperity across Europe leading to better housing and social conditions as well as advances in medical treatments and eradication of many infectious diseases, such as TB and Polio, which have historically accounted for many premature deaths. This growing elderly population is increasingly more healthy and active and now forms a substantial consumer niche in its own right. Figures 2.2 and 2.3 show how Polk Ltd portrays these markets in the niche lifestyle marketing categories.

The implication for retailers is quite profound. As Goodwin and McElwee (1999) note, age is important not only for what is bought and sold but also how products are sold. There will be a relative decline in demand for products and services particularly consumed by young adults. These are likely to include items such as fashion, leisure, entertainment and sports-related products. On the other hand, more affluent middle-to-later-aged adults will increase the demand for financial services products, such as investments and pensions, holidays, home furniture and appliances. According to Field (1997), the 'grey' market has an income 8% higher on average than the rest of the population. In the financial services market, this means that 90% of them have savings and investment plans, 52% save regularly, 43% have a 'Tax Exempt Special Savings Account' (TESSA) account and 38% have a 'Personal Equity Plan' (PEP) (Field 1997). Assuming zero growth in expenditure, we can expect some of these retail sectors to grow or decline broadly in line with the age cohort shifts. In terms of how goods are sold, it seems apparent that certain types of (new) channel may be more suited to certain age groups (see Chapters 3 and 5).

2.2.5 Changing household size

Throughout Europe the average size of households is decreasing at quite a rapid rate. Table 2.6 illustrates how average household size has decreased in a number of European countries over the past 10 years and how projections suggest that this will change in the future, while Table 2.7 shows the more detailed change within the United Kingdom. The reasons for the reduction in average household size are numerous, and include a range of lifestyle factors. Perhaps the most important are the increasing divorce rates across Europe, wherein the traditional nuclear family, once the norm, is now becoming the exception. In fact, only about 25% of all UK households in 1998/99 consisted of a 'traditional' married family with children, while around 30% of all households now are classed as 'non-family' households (including single, widowed and divorced persons as

Easy street

My husband Harold is 65 years young. Oh, he could retire, but he happens to be a very successful businessman and he enjoys his work. Besides, he wouldn't want to tag along with me on my afternoon socials and he just doesn't care for shopping like I do. I have a rule, you know ... never had cash.
I put everything I buy on my credit cards. I think that credit cards are the greatest thing ever invented, next to those little individual pudding cups of course.

Anyway, Harold does take lots of time off, especially when our son and his wife come to stay at our home. They taught us how to use our computer. I never touch the thing myself but Harold bought it to keep our finances in order. I've always said I married him for his mind, but he also has a pretty good line... of credit, that is.

Figure 2.2 Niche profile – 'Easy Street'

well as co-habiting non-married couples). Also, more people are delaying marriage until later in life, creating a demand for more single households between ages 18 to 35. The growth in single person households (see Table 2.7) is the biggest cause of the fall in average household size.

From a retailing perspective, the implications of these changes are interesting. First, with average household size falling there is a natural increase in the number of dwelling units required to house a static population. In the United Kingdom, Government estimates suggest that we now need 3.8 million new households (1999–2021) to accommodate these changes in lifestyle (and of course there is an interesting geographical question as to where these may be located). This suggests that there will be a growth in the demand for smaller residential units. Since every new household has to be furnished and stocked with basic goods, this suggests that there will be an increase in demand for household appliances and furnishings as well as other residential services such as heating, telephone, and so on.

When my brother and I were born, my grandma and grandpa gave us each a savings bond. They like to buy those kinds of things. Then I get money for the bond sometime when I get older. Cool, huh? So maybe then I'll have enough saved to travel like grandma and grandpa do all the time. I like to look at all their pictures of places like Italy and France and Australia.

I also like to help grandma in the garden when I visit. She plants all sorts of stuff, not just flowers, but tomatoes and carrots and broccoli. It's fun to help her plant it though and watch everything grow.
We're going to have a big party for grandpa's 70th birthday next month. I helped grandma pick out presents for him that you send for through the mail. Grandma loves to do that 'cause she says it's so much fun waiting for the stuff to arrive.

Figure 2.3   Niche profile – 'Nomadic Grandparents'

2.2.6   Population growth and redistribution

Although the rate of population growth within most of the developed world is relatively modest, there are some major changes that will take place at the national and regional level. At the national level in Europe, for example, population growth is currently highest in Luxembourg, Ireland, Netherlands and France, and lowest in the Scandinavian countries. Within a country, population change can be far more dramatic. In the United Kingdom, for example, we have witnessed population decline in areas such as Manchester, Liverpool and the West Midlands, while counties such as Essex, Northamptonshire, Hampshire, Cambridgeshire and North Yorkshire have seen population growth as high as 20% over the last decade. There is a dearth of good population forecasts below the regional level. However, some consultancy organizations, such as Experian, CACI and GMAP, do offer

Table 2.6 Change in occupants per household by EU member state 1977–1996

Austria	2.86	2.47
Belgium	2.97	2.70
Denmark	2.53	2.20
Finland	2.80	2.38
France	2.91	2.55
Germany	2.58	2.32
Greece	3.46	2.88
Ireland	4.09	4.10
Italy	3.25	2.57
Luxembourg	2.93	2.94
Netherlands	2.96	2.43
Portugal	3.42	2.74
Spain	2.90	2.63
Sweden	2.40	2.27
UK	2.87	2.70

Source: Eurostat.

Table 2.7(a) Declining size of households in the UK, 1951–2011

	1951	1961	1971	1981	1991	2001	2011
Average household size	3.45	3.2	2.8	2.7	2.4	2.3	2.25
Number of households (m)	15	17	17.5	19	22	24	26

Source: OPCS.

Table 2.7(b) Changing household types in England, 1981–2011 (m)

	1981	1991	2001	2011
Married couple	11012	10547	10217	10037
Co-habiting couple	500	1222	1447	1549
Lone parent	626	981	1202	1259
Other multiperson	1235	1350	1671	2051
One person	3932	5115	6509	7875
All households	17306	19215	21046	22769

Source: Department of Environment, UK.

population forecasts as part of their client services. Figure 2.4 shows Experian's forecasts for population growth in the United Kingdom Yorkshire and Humberside region (for more discussion on forecasting methods and small-area predictions, see Debenham *et al.* 2001).

For retailers, there is a need to assess the implications of these changes in population geography, particularly on the location of outlets. For example, if the bulk of a retailer's network is located in areas with small or negative projected growth, then there could be serious implications for long-term viability. Conversely, retailers need to ensure that they have an appropriate strategy for exploiting the growth in the various dynamic and

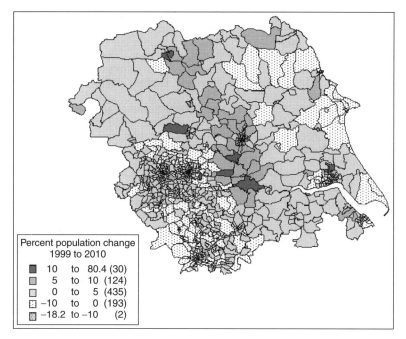

Percent population change
1999 to 2010

10	to 80.4	(30)
5	to 10	(124)
0	to 5	(435)
−10	to 0	(193)
−18.2	to −10	(2)

Figure 2.4 Estimated population change 1998–2010: Yorks & Humbs. Source: Experian

prosperous regions of both the countries they are located within, and those that they may wish to be located within in the future.

2.2.7 New consumer groups

The marketing world is constantly on the look out for niche market segments, and it seems to be, that from time to time, new consumer groups appear to become important (or more important). We have already alluded to the growth of the 'grey' market in Section 2.2.4. Croft (1994) provides a very useful summary of the types of market segmentation common in the retail and marketing world. Some of these have always been important: for example, income. In the United Kingdom, Sainsbury's and Tesco in the grocery market have always aimed more at middle and higher income customers, leaving the discount market to retailers such as Kwik Save and the Co-op (with Aldi and Netto exploiting the deep discount sector in more recent times: see Section 3.4).

The traditional age, sex, social class segmentation of the retail marketing literature has now been extended to include many of the other large consumer groups. Ethnicity has always differentiated consumers, but retailers have often been slow to tap into such differences. Understanding the variations in consumption patterns among different ethnic groups has also been a focus of recent academic study. Lamont and Molnar (2001), for example, take an interesting look at Black American consumers. They explain the very different black consumption patterns (far more spent on personal care services, clothes, footwear and expensive sports cars) on the desire to show their social identity and to make a statement to non-Blacks that they too can spend money on conspicuous consumption, and hence, create a positive vision of their cultural distinctiveness. Other new groups include the gay market. A number of retailers have already jumped on the gay bandwagon, with companies such as Ikea, Virgin and American Express having images of being more customer-friendly to gay consumers (Field 1997).

The so-called 'green market' has also risen dramatically over recent years. This builds on the desire for more ethical forms of production and consumption, especially organic methods of farming. Increasingly, consumers are concerned about food quality and many blame the dioxins in food for recent health scares. Confidence in processed foods, in particular, is currently very low (Dawson 2000). It also reflects a growing concern over beauty products tested on animals in laboratories. Much has been made in the literature on the growth of companies such as Body shop and their entrepreneurial politics of 'profits with principles' (Kaplan 1995). A number of new retailers are also exploiting the greens. One of the fastest growing food chains in the United States is Pratt's Wellmarket, a new chain of organic grocery retail stores, while in the United Kingdom, Boots has increased its number of 'Pure Beauty Cosmetic' shops (for a more general discussion of the rise of the green market see Wagner 1997).

2.3   Trends in Distribution and Retail Supply

2.3.1   Consolidation: concentration of market power

The late twentieth century has witnessed the emergence of a set of superpowers controlling retail distribution. Figure 2.5 shows the market value of the leading organizations in 1999. Essentially, a new world order is developing as these global giants look to dominate the retail landscape. By 2010, we can expect a smaller number of larger companies to have emerged, encompassing a diverse range of retail formats and propositions. Why has this concentration of power occurred?

Most detailed work to date on retail consolidation has concentrated on the grocery market, and we can draw on this material to provide a useful case study. Although consolidation has occurred throughout the twentieth century, many analysts point to the 1960s as the decade of major change in most western countries. In the United Kingdom, for example, 1963 is seen as a key year when retail price maintenance (RPM) was abolished, enabling major grocery retailers to pass on cheaper prices to the consumer. Thus, following the abolishment of RPM, price became the most important competitive attribute in the grocery industry. Retailers 'piled it high and sold it cheap' and the customers left the independents in droves. Price wars became the norm until the end of the 1970s, when the focus shifted away from price towards quality. At the same time, retailers recognized that growth could be gained by controlling costs and improving efficiency.

One key way to control costs was to gain greater economies of scale through bulk buying and squeezing manufacturers to offer better and more exclusive deals. The battle for power between the manufacturing and retailing sectors is well discussed by Crewe and Davenport (1992), Foord *et al.* (1996) Hughes (1996, 1999), Duke (1998) and Ogbonna

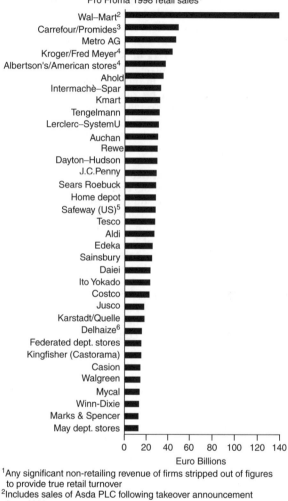

World's top 35 retailers, mid-1999, ranked by
Pro Froma 1998 retail sales[1]

[1]Any significant non-retailing revenue of firms stripped out of figures
 to provide true retail turnover
[2]Includes sales of Asda PLC following takeover announcement
 June 1999
[3]Following merger announcement August 1999
[4]Mergers completed May 1999 (Kroger), July (Albertson's).
[5]Includes sales aquisitions: Dorminick's, Carr Gottstein and
 Randall's
[6]Includes sales of Delhaize American companies: Food Lion,
 Kash n'Karry and Hannaford   (aquisition announced
 August 1999)

Figure 2.5   World's top retailers by sales. Source: Wrigley (2000)

and Wilkinson (1998). Basically, there was a shift during the 1970s and 1980s from the manufacturers dictating price and conditions of sale to the retailers calling the tune. The power of the retailer was heightened by their ability to introduce substantial savings by reducing the circulation costs of capital – that is, by only paying for goods once they had been sold, thus putting more of the cost of stock holding on to the manufacturers.

At the same time, leading grocery retailers were transforming retail distribution and cutting costs through greater use of own brands, labour force restructuring and improvements in IT and distribution. The major aspect of branding was the growth in own labels, where even more stringent requirements could be placed on manufacturers (although today branding refers to much more than simply retailing own brands – see e.g. Laaksonen and Reynolds 1994). By the end of the 1980s, over half of the sales of Tesco and Sainsbury's came from own brands (Ogbonna and Wilkinson 1998: see also Hughes 1996, 1999). In terms of labour productivity, the employment of greater number of female part-time workers was part of a desire to enhance labour productivity and reduce labour costs (see Shackleton 1998). Investment in IT came in two major ways. First, great efficiencies were made in distribution through major new depots and service vehicles (see Section 4.6). Second, retailers developed new ways of capturing sales information through electronic point of sales data (EPOS). This enabled retailers to achieve operational control by centralizing their activities from purchasing to distribution (Clarke I. 2000). In particular, just-in-time systems allowed the immediate restocking of goods when, and only when, the sales data had been recorded via EPOS at the store. This sales data also enabled retailers to generate accurate data on customer buying habits that were used to refine the marketing mix (Clarke I. 2000).

However, the manifestation of the growth of the leading grocery retailers mostly came from the new stores they built during the 1970s and 1980s, continuing into the 1990s, albeit at a slower rate. Wrigley (1989, 1994) described this period as the 'golden era of supermarket retailing', as new 'cathedrals of consumption' appeared at the edge of our cities. Wrigley has also used the term *store wars* to capture the essence of a battle that was largely about getting the best sites in order to maximize sales. It was obviously a geographical battleground, as each retailer tried to increase market share away from its heartland. Indeed, during this period many of the top 10 retailers emerged as national rather than regional players.

Despite a slowing down of growth in the 1990s (caused by tighter planning controls, asset depreciations and greater competition from the discount sector: Wrigley 1994, 1996), by 2001, the top companies held between 50 and 80% in most western countries. Interesting case studies of retail grocery change in particular cities can be found in Guy (1996b) and Clarke I et al. (1994)

It should be emphasized that the consolidation of power in modern retailing has claimed many casualties among leading retailers and among small independents and multiple groups. Companies that trade well and often achieve spectacular success are prime targets for takeover bids (often aggressive rather than friendly bids). Managing fast growth is not easy. Indeed, Dawson (2000) suggests that the biggest challenge retailers face today is the issue of how big to grow, and how to manage that growth. The methods described in Chapters 6 to 10 may go some way to aiding that management process.

The geography of this corporate power battle has also been fascinating to witness. Having secured high domestic market shares, most of the leading elite group of retailers (mainly in the United Systems and Europe) looked to even more rapid expansion in overseas markets, especially the emerging markets of Latin America, Asia, Central and Eastern Europe and Southern Europe. This development is outlined in more detail in the next Section.

2.3.2 Retail internationalization

Although by no means a new phenomenon (see e.g. Whysall 1997), retailing has increasingly taken on an international dimension, with many organizations expanding their operations outside their traditional domestic market. Tables 2.8 and 2.9 show the international operations of two of the biggest world retailers in 2000, Wal-Mart (US) and Carrefour (France), respectively.

Alexander N. (1997) has identified five main reasons for retail organizations focussing on expanding their international operations. For each of these factors, both push and pull effects are at work. The first factor is politics. The push factors here would include

Table 2.8 Wal-Mart international operations, January 2001

Country[a]	Year of entry	Method	Format	Number of stores	Floorspace (mn ft^2)
Argentina	1995	Wholly owned	Supercentre	11	2.07
Brazil	1995	Majority owned (was joint venture)	Supercentre	12	
			Sam's Club	8	
				20	3.04
Canada	1994	Wholly owned	Wal-Mart Stores	174	20.48
China	1996	Joint venture	Supercentre	10	
			Sam's Club	1	
				11	1.65
Germany	1997/98	Wholly owned	Hypermarkets	94	9.20
Korea	1998	Wholly owned	Supercentre	6	0.85
Mexico	1991	Majority owned	Supercentre	32	
			Sam's Club	38	
			Other	429	
				499	22.03
Puerto Rico	1992	Wholly owned	Wal-Mart Stores	9	
			Sam's Club	6	
				15	1.78
United Kingdom	1999	Wholly owned	ASDA Stores	238	
			ASDA Supercentre	3	
				241	19.28
Total				1071	80.38

[a]Wal-Mart also had stores in Indonesia (1996) but these have been closed and there was an earlier joint venture into Hong Kong (1994), which was ended.
Source: Wal-Mart 10-K reports to the US Securities and Exchange Commission, *http//www.sec.gov*, Burt and Sparks (2001).

Table 2.9 Carrefour's International Interests 2001

Country	1st Opening	Stores
France	1963	3360
Spain	1973	2711
Greece	1991	351
Portugal	1991	312
Turkey	1993	70
Italy	1993	878
Poland	1997	55
Belgium	2000	492
Czech Republic	1998	7
Slovakia	2000	2
Switzerland	2001	11
Brazil	1975	199
Argentina	1982	377
Mexico	1994	19
Colombia	1998	3
Chile	1998	3
Taiwan	1989	24
Malaysia	1994	6
China	1995	27
Thailand	1996	13
S.Korea	1996	21
Singapore	1997	1
Indonesia	1998	7
Japan	2000	3

an unstable political structure within an organization's domestic markets coupled with a restrictive regulatory environment, and also perhaps, restrictions on consumer credit. The pull factors are essentially the mirror image of these push factors – a stable political structure and a relaxed regulatory environment with a pro-business culture. The dynamics of the latter largely accounts for the interest of Sainsbury's and Ahold in the US market (see Section 2.4 for more on the importance of the regulatory environment).

The second set of factors is economic. Poor economic and trading conditions within a domestic market (along with maturity or perceived saturation in a particular market sector) will act as push factors for retailers looking to expand beyond their domestic market place. Countries that have good economic growth prospects, low operating costs and growing markets, will be targeted as development opportunities for retailers. In a European context, countries such as Poland, the Czech Republic, Slovakia and Slovenia all fall within this category. On a global scale, countries in south East Asia have had (until recently) the same characteristics and have attracted major investment from United States and European retailers. At this point, a case study is useful to illustrate this important factor.

Tesco's expansions into Central and East Europe are now well advanced. In 2001, they operated 45 stores in Hungary and 40 in Poland. Newer ventures in the Czech republic and Slovakia occurred in 1996. This created an operating environment on the basis of four contiguous East European countries, which provides significant savings in distribution. It is useful to provide a little more background on these developments. In 1994, Tesco invested in the Hungarian retailer, Global, and opened the first store under the Tesco name in the town of Szambathely in 1995. In November 1995, Tesco invested £8 million in Savia, a small food retailer in Southern Poland. In March 1996, they agreed to buy the two retailing businesses of the US retailer, Kmart in the Czech Republic and Slovakia for £77 million. Kmart's business consisted of 13 food and general merchandise stores in the main cities of those two countries. In 1995, sales in Hungary and Poland amounted to £45 million. Since entering the Central European market place, Tesco has focussed on the development of large hypermarkets of approximately 100 000 sq. ft of floorspace. They sell a wider range of non-food products than their UK stores – the whole operation being geared towards one-stop shopping. In 2000 alone, Tesco opened six new hypermarkets in Hungary, six in Poland, two in Czech Republic and three in Slovakia, adding well over one million sq.ft.

So what are Tesco's reasons for developing this Central European business? Central Europe offers a number of advantages from a supermarket perspective. First, levels of competition are relatively low, with few big groups emerging to generate the kind of discounting wars that we have witnessed in Western Europe. Secondly, the opportunities for growth are substantial – planning regulations and control are currently more lax than elsewhere in Europe, allowing Tesco to roll out its hypermarket format at some pace. Thirdly, Central Europe is enjoying a period of economic growth, following the collapse of communism and the opening up of their markets. Finally, because Tesco operate in four adjoining countries, they can develop cross-border logistics and distribution operations, thus generating economies of scale.

The expansion plans for Tesco are not limited to central and Eastern Europe, and other markets are now seen to be equally important in economic terms. By 2000, Tesco had opened 24 stores in Thailand, 7 in S.Korea and 1 in Taiwan (in 1999, Tesco announced a joint venture with Samsung to develop superstore sites in South Korea). Indeed, Asia looks set to be a fascinating battleground for western retailers, provided the economy of Asia recovers from the problems of the late 1990s (Dawson 2000). Markets such as China, for example, offer huge potential. Currently, only 3 to 4% of the 7000 supermarkets currently in operation in China are in the hands of foreign retailers, and therefore, offer great potential for retailers such as Wal-Mart and Carrefour (Goldman 2000). It should also be noted that the attractions of such large economic markets are apparent outside the grocery industry. The UK DIY retailer B&Q, for example, recently announced a major store expansion programme in China.

The third set of factors used to explain growing internationalization is purely demographic. Some countries in Europe, such as Germany, Greece, Italy and Spain are experiencing virtually zero or negative population growth, and therefore, present little opportunity for growth in some of the retail sectors. The growth of populations in China and the Far East only add to the interest in these markets from western retailers.

Fourth, there are cultural factors. Cultural similarities between countries are often deemed to encourage internationalization. A common language is perhaps the clearest manifestation of cultural similarities, and it is therefore not surprising that US retailers have seen the United Kingdom, Canada and Australia as natural markets for international expansion. Similarly, UK companies have seen countries of the Commonwealth and the

United States as natural markets for overseas expansion, despite the proximity of much larger markets within continental Europe. A common language not only assists in managing the business but also in marketing and branding. However, sharing the same language can also hide deeper cultural differences, as Winston Churchill once famously said, 'the United Kingdom and United States were two countries divided by a common language'. A good case study of potential pitfalls is provided by O'Grady and Lane (1997). They examined the performance of Canadian retailers in the United States. Although geographic neighbours, enjoying similar qualities of life and a common language, Canadian executives were reported to have found significant differences in cultural terms. The lack of prior recognition of such differences caused a number of high profile market entry failures. Indeed, they argue that of the 32 main entries into the US market, only 7 have achieved profitable success. Similar arguments have been put forward to account for Marks and Spencer's problems in Canada and the failure of the Early Learning Centre in the United States.

There are many examples of organizations that have struggled to be successful abroad because of fundamental differences in retail culture, especially when there is no common language. Ford, for example, pulled out of Russia in the 1920s because of cultural problems. Attempts by Ford to build up custom in the Former Soviet Union (FSU) countries in the mid-to-late 1990s are still bedevilled by cultural problems, the most important of which is the method of doing business (Whittam and Clarke 2002). Some FSU countries are still plagued with red tape and corruption. Some organizations may have to accept bribery as a prerequisite of business and good intent (Alexander N. 1997, Whittam and Clarke 2002).

Although cultures can be very different, there is an interesting debate as to whether there is a growing 'Euroconsumer'. The seemingly universal demand for fast food is often quoted as a good example, and is said to explain the success of firms such as MacDonalds. However, most European-wide organizations still tailor their products to local markets, even MacDonalds. Lynch (1993) shows the variation in product range of MacDonalds across Europe. Benetton, too, are widely known to vary their product range extensively in Europe. These trends lead many analysts to question the notion of the Euroconsumer:

> There is no such thing as a standard Euroconsumer and so there is no such thing as standardized Euro-retailing. The attempts to explore the presence of common consumer lifestyles across Europe (Cathelat 1991, Sampson 1992) only serve to illustrate the variety of consumers as much as they point to convergence of attitudes and behaviours. (Dawson 1995 p.7) – see also Dawson (2000) for a reiteration of this argument.

However, Robertet (1997) has attempted to classify European consumers into six major categories (see Table 2.10). Although all six are present in most countries, each is more prevalent in different parts of Europe. Thus, the discount shopper is most common in Germany, the mall shopper in France, and the neighbourhood shopper in Italy. Guy (1998b) is more skeptical. He argues that the variations in the distribution of these categories may reflect the retail geographies of these countries as much as any cultural stereotypes. However, as he has also noted, whether the retail pattern influences the consumer characteristics or vice versa, is hard to determine.

Alexander's fifth explanation for internationalization is the retail infrastructure itself. Any hostile and competitive trading environment with high concentration levels may

Table 2.10 Six types of European consumer

Type	Percentage of European population	Description
Smart shopper	13	Young urban people, well educated, like novelty. No preferences for type of shop, will seek good value for money.
Demanding shopper	13	Older people, well-off and selective. Seek good standards of service and are loyal to specialist shops.
Mall shopper	12	Suburban, well-off people who prefer to obtain everything under one roof, either in hypermarkets or shopping malls.
Old neighbourhood shopper	21	Older people less well-off. Conservative, value tradition and their local roots. Fixed shopping habits, favour the 'corner shop'.
Materialistic bargain hunter	11	Mainly blue collar, are avid shoppers and will use any type of outlet, but will never pay full price for anything.
Discount shopper	30	Tend to be lower income, of all age groups and types of locality. Prefer shops which are local and offer lowest prices.

Source: Based on Robertet (1997). Reprinted in Guy (1998b).

encourage retailers to seek expansion outside their domestic market. This is often the reason many retailers cite for their moves overseas – domestic market saturation (see e.g. Alexander and Mortlock 1992). We shall explore the notion of saturation in much more depth in Section 2.3.3.

To Alexander's list could also be added the importance of new, emerging global centres. In the high fashion sector, for example, premium brands are targeting global locations such as Bond and Oxford Street, in London, where they are selling premium products at premium prices. Fernie *et al.* (1997) describe the growth of these top brands in the West End of London. The impact of this trend is to push up rents so that other forms of retailers have had to vacate this territory and move to less expensive retail space. Indeed, Oxford Street is now the third most expensive retail location anywhere in the world (behind other major fashion locations such as Madison Avenue in New York and Causeway Bay in Hong Kong).

There is a growing literature on international retailing and the interested reader is encouraged to look at Treadgold (1991), Dawson (1993), McGoldrick and Davies (1995), Alexander N. (1997), Lamey (1997) Sternquist (1998).

2.3.3 Reduction of costs

It is clear that during the last decade retailers have become much more conscious of reducing waste and cutting costs as retail competition has hotted up. Two case studies will be useful for illustrating this trend. First, the retail financial services industry in Europe is experiencing a massive transformation that is reflected in its diminishing physical representation within the retail landscape. This follows many years of growth and expansion. During a period of industry deregulation and consumer spending boom in the late 1970s and 1980s, the financial services market experienced strong growth as a direct result of greater consumer participation. According to Leyshon and Thrift (1995), the level of personal debt in the United Kingdom more than doubled between 1980 and 1993, from £23.1 billion to £50.7 billion pounds, as consumers expressed confidence in their ability to finance long-term debt. This period of industry growth was characterized by an increase in the size of the branch network for many financial service organizations. In particular, those organizations that operated relatively small branch networks found it advantageous to rapidly expand their delivery network by opening branches in smaller suburban centres or small towns and in close proximity to the many new housing developments. As financial service organizations found themselves thrust into new, highly competitive markets (with supermarkets, in particular, gaining considerable market share), there was an urgent need to adjust operations to become more effective and efficient. This requirement gave rise to two key cost-cutting strategies affecting branch location: the automation of consumer services and the reduction of the delivery network.

In the United Kingdom, approximately 2400 branches or 20% of all outlets were closed by the major banks between 1989 and 1994. Most of the large established banks that were involved in branch closures viewed their branch networks as an inefficient way of meeting customer needs within the changing market (Trefson 1994). For the most part, the closure strategy was not terribly sophisticated as the smallest, least profitable branches, perceived to be servicing the lower value consumer areas, were closed first. Leyshon and Thrift (1995) report that these branches were often the smaller suburban branches opened in the boom periods of the 1980s (especially in inner city locations). Further, in 1996, National Westminster Bank announced plans to close 200 branches by 2001.

The period of rationalization has also coincided with a huge programme of mergers and acquisitions over the last decade, which has resulted in a smaller number of larger players (see Section 2.3.1 and Section 4.3). This allowed the new consolidated banks to prune their branch network as redundant branches created through the merger process were closed. However, there are still enormous variations in the number of bank branches per capita across Europe, as shown in Table 2.11. This table suggests that countries such

Table 2.11 Rates of branch provision

Number of bank branches		Number of branches per 10000 of population		Average number of staff per branch	
Belgium	10511	Belgium	10.4	Belgium	8
France	25317	France	4.4	France	12
Germany	50206	Germany	6.12	Germany	20
Italy	22391	Italy	3.94	Italy	20
Netherlands	5211	Netherlands	3.41	Netherlands	14
Spain	35240	Spain	8.99	Spain	8

as Spain and Belgium still have a long way to go in terms of rationalization to meet the levels of provision currently experienced in more mature markets. It is suggested that even in the markets that have shown most rationalization, bank branch provision is likely to decrease dramatically in the future.

A second example is provided by the car industry. Until recently, there has been a comfortable, symbiotic relationship between manufacturers and dealers within the industry. However, owing to increasing competition between manufacturers, retailer margins have declined substantially. This has put substantial pressure on the reduction of distribution costs within the system. As a consequence, in most European countries, there has been a substantial reduction in dealer numbers. However, there remain substantial variations in the number of dealers per thousand population within different European markets, as exemplified by Table 2.12. It is difficult to comprehend that there are more car dealers in Germany than there are in the United States, a country with a car market that is about three or four times the size of Germany's.

One of the causes of reduced margins is the level of intra-brand/inter-dealer competition, particularly within metro markets. For example, in a typical metro market such as Leeds, Dortmund, Barcelona or Naples, there are historically three or four appointed dealers with their own franchise territory. However, from the consumer point of view, these defined territories are irrelevant, and the close proximity of a number of competing dealers allows consumers to trade one or more dealer propositions against another. This has the net effect of competing away the dealer profit associated with the new car purchase, and of consumers ultimately buying on price rather than any other factor. The response from many major manufacturers has been to define larger (customer marketing area) territories, allocating a single dealer to cover a whole metropolitan market area. In principle, this should lead to the retention of a greater proportion of the margin as the level of inter-grand/intra-dealer competition is removed.

Cost cutting is also an exercise that appeals to manufacturers and in some sectors, this has had profound impacts on retailers. As seen in Section 2.3.1, within the general retail environment, there has been a continual battle between manufacturers and retailers. The balance between the two has varied, for example, between dominance of the supermarket chains over manufacturers, and in contrast, the control that car manufacturers exert over their dealer networks. However, in many markets, the retailers seem to have the upper hand. To fight back, many manufacturers are now considering the process of 'disintermediation'. This refers to the elimination of layers of added cost from the distribution network by companies that have previously relied on other organizations to sell their products. The

Table 2.12 Dealers per 100 000 population

Switzerland	56	Netherlands	24
Belgium	51	Ireland	21
Austria	45	Finland	20
Norway	43	Sweden	19
France	39	UK	14
Germany	36	Portugal	11
Denmark	33	Italy	10
Spain	27		

growing number of manufacturers engaging in retailing includes Reebok, Nike, Levis and Daewoo. This has especially been the case in the fashion industry as we have witnessed the emergence of global retail brands such as GAP, Levis and Nike. These brands have built their presence around massive corporate advertising initiatives, including sponsorship and endorsement from major sporting and media personalities. Increasingly, these brands are available through their own retail outlets. Nike, for example, have created massive consumer interest in their Nike Town formats that they have rolled out in major retail locations such as Chicago, Los Angeles and New York, which are increasingly looked upon as crucial centres of world importance for the fashion business. Here, they devote considerable amount of floorspace to individual sports such as baseball, basket ball, soccer, golf, etc., and these outlets serve as major retail destinations even for consumers who can purchase Nike products in more convenient locations. Similarly, service providers such as British Airways and Warner Brothers are now involved directly in store operations. Field (1997) notes that the economics of direct selling are seductive with margins of 18 to 25%. Similarly, in the car industry, between 20 and 30% of the showroom cost of a vehicle is accounted for by post-factory gate costs, including marketing and dealer margins. It is thus not surprising that we are beginning to see manufacturers selling cars direct, as in the case of Daewoo in Europe. They now have 150 outlets in their own UK retail network and, additionally, retail from computer links in other retail stores (notably Halfords). These costs clearly have to be outweighed against the benefits of selling through third parties.

Another major component of the rationalization activities described above comes from sections of the industry that are convinced that home market saturation has occurred. These are often the sectors where, after many years of operation, mature markets are left with a few major players enjoying high sector market shares. Thus, there is a common belief that few (if any) new viable sites remain and that future increases in profits will have to come from cost savings or internationalization, rather than from revenues from new domestic stores. This fear of saturation is important enough to be considered as a major trend in the retail industry.

Away from the grocery market, Shell suggested in 1997 that the petrol market had reached saturation, especially with the increase in petrol retailing by the supermarkets. As we also saw in the preceding text, most financial service organizations believe there are too many high street banks, especially given the rise of new forms of distribution channel (see Chapter 3 and 5).

The problem with the saturation thesis is that markets are often deemed to be saturated, without any real analysis of what saturation actually means. Too often, retailers are advised by consultants that no growth can possibly be obtained in the saturated markets of West Europe. For example, following a pilot study of variations in market potential undertaken at the University of Leeds in the early 1990s, the then chairman of W. H. Smith (the UK book and record retailer) expressed surprise that so many new regional opportunities were apparent for his company after previous consultancy reports had talked of nationally saturated markets and no further potential for geographical growth. Indeed, W. H. Smith announced in 2001 that it intended to find sites for another 100 stores to be added to its network.

The notion of saturation is a controversial one as much of the debate on saturation (and resulting internationalization) takes place without any spatial analysis of variations within and between different market sectors. The history of work in the grocery market provides a good illustration of the difficulty associated with this concept. The possibility of grocery market saturation in the United Kingdom was first mooted by Jones P. (1982). Since then we have seen the 'store wars' battle for market share described in Section 2.3.1 and a consequent huge increase in superstore/hypermarket floorspace, although part of this new floorspace actually replaced smaller stores that were closed in traditional high streets, a point picked up by Myers (1993). Another wave of concern over saturation appeared in the late 1980s and early 1990s. Duke (1989) again raised the issue of saturation in the grocery market, but again, without any real recourse to analysis. At the same time, senior executives in the grocery industry pronounced that saturation was here or was very close. The Tesco Chairman Ian Maclaurin talked about superstore 'finality' in 1987 while Archie Norman, the then CEO of ASDA, claimed in 1992 that saturation was just five years away.

The first major spatial examination of saturation came from Myers (1993) when he looked at regional variations in levels of provision. He was the first to show that great variations in provision existed across the United Kingdom, and that the notion of saturation seemed erroneous. Guy (1994a, 1996a) similarly argued that saturation could only be considered in a local context in his two papers on retail change and consolidation in Cardiff. However, the notion of a finite level of provision continued to be raised in some quarters of the literature. For example, Duke (1998) claimed:

> UK grocery retailers currently stand at a watershed: massive programmes of site acquisition and store building are virtually defunct, killed by saturation, the near exhaustion of viable superstore sites (p. 93).

Langston et al. (1997, 1998) were the first authors to look in detail at regional and local provision in the mid-to-late 1990s as they built on the work of Myers (1993). They too argued that saturation can only ever be a local phenomenon – that is, it is unlikely that provision is ever uniform over an entire country or region. Similarly, Lord (2000) poses the interesting question of whether saturation is simply 'a moving target that is never reached because of the self-cleansing nature of the retail sector, the introduction of new formats, improvements in the efficiency of retail operations and shifts in consumer demand' (p. 342). Langston et al. showed the variations between regional and local provision rates in the UK grocery market to be large. This analysis has since been updated by Poole et al. (2002b). Figure 2.6 shows the variations in floorspace per head of population in the UK grocery market in 2001. It can be seen that there are huge variations from the high floorspace levels per capita in and around outer London to the low levels in parts of Midlands and Northern England. Even in the most 'saturated' counties such as Cheshire, there are still widespread intra-county variations (see Poole et al. 2002a, 2002b for more details).

In addition to showing that spatial variations in floorspace provision per head of population still exist and are significant across the United Kingdom, Langston et al. (1998) showed how new store growth continued, thus arguing that saturation looked far from imminent. This theme is taken up again in Section 3.3.

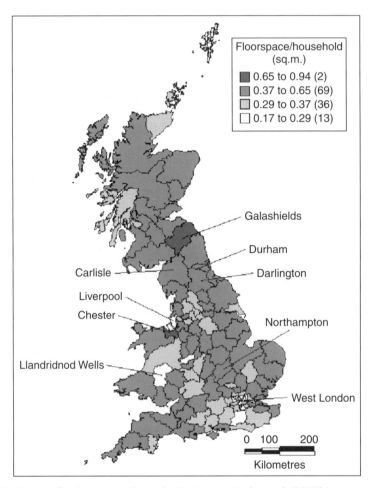

Figure 2.6 Grocery floorspace per household. Source: Poole *et al.* (2002b)

The same historical concerns over saturation have been seen in many other markets of the world. Cliquet (2000a) notes that at the beginning of the 1970s a Nielsen study showed that between 200 and 300 hypermarkets were all that were necessary to saturate the French market. By 1999, there were more than 1100 hypermarkets in France. Poole *et al.* (2002a) show similar patterns of variation to the UK market for the French and German grocery retailing markets. Figure 2.7 shows the variations in provision per head across France, where northern and western areas seem far better served than southern

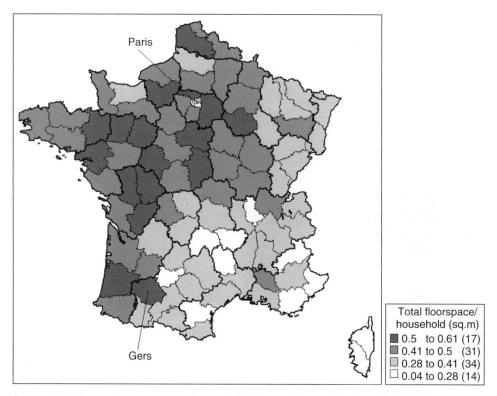

Figure 2.7 Grocery floorspace per household, France. Source: Poole *et al.* (2002a)

and eastern areas. Figure 2.8 shows that provision rates are highest in the far west, north and south of Germany.

Work has also been carried out on provision rates in the United States. Silcock *et al.* (1999) show the variation in provision levels per head of population across the United States (see Figure 2.9). As with the European examples in the preceding text, Figure 2.9 shows that many of the largest urban areas have the lowest provision rates relative to more rural areas. This, at first glance, seems counter-intuitive. However, the fact is that many smaller towns and rural areas have the space to build large stores and are often saturated by power retailers such as Wal-Mart. The larger cities have more floorspace in absolute terms, but less per head of population than their more rural counterparts.

O'Kelly (2001) looks at saturation at the level of the organization. He examines the relationship between market share and share of space by retailer in a set of US urban areas. This focus on individual retailers is useful. Bennison *et al.* (1995) take a similar

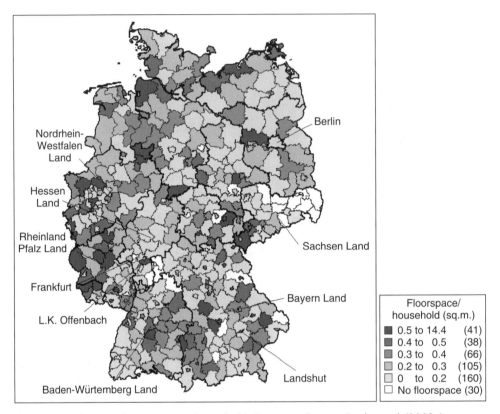

Figure 2.8 Grocery floorspace per household, Germany. Source: Poole *et al.* (2002a)

approach and argue that market saturation can be identified where additional stores do not generate proportional increases in firm sales.

The important argument in the saturation debate is that if retailers can make good profits in areas of high provision, then there must be opportunities in other areas for future growth. Exploring local levels of saturation helps to reveal potential new opportunities. If organizations can also work out their local market shares in these less saturated areas then they can identify areas of both low overall provision and gaps in their existing network. They may then be able to identify new locations for organic growth (see Chapter 3). True saturation can only occur when excess profits are eliminated altogether. We argue that this is one area of future research that should reap many benefits. In addition to plotting indicators of provision, it is useful to also investigate where potential battlegrounds lie. A key indicator may well be the percentage of the market not in the hands of major players. Figure 2.10 plots this indicator for the US grocery market.

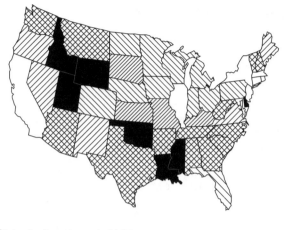

States by floor_household 92

■ 12.1–16.2 (7)
▨ 10.5–12.1 (14)
▨ 9.4–10.5 (8)
▨ 8–9.4 (14)
▢ 3.7–8 (8)

Figure 2.9 Grocery floorspace per household, United States. Source: Silcock *et al.* (1999)

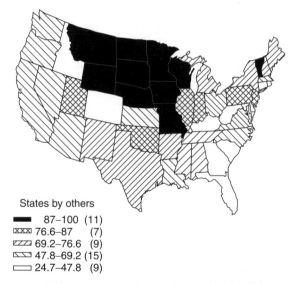

States by others

■ 87–100 (11)
▨ 76.6–87 (7)
▨ 69.2–76.6 (9)
▨ 47.8–69.2 (15)
▢ 24.7–47.8 (9)

Figure 2.10 Percentage of US grocery market not in the hands of the top 10 retailers 1997. Source: Silcock *et al.* (1999)

2.4   The Regulatory Environment

In Section 2.3.1, we outlined the conditions that led to the growth of superstore retailing. Most of these new stores were away from city centres, and often in green-field, edge of city sites. However, there are still considerable variations in the importance of large grocery multiple retailers between different developed countries, and hence, the rate of concentration has not been even geographically. A crucial factor in the speed of development has been the variable degree of local or national planning legislation. Most planning regimes in Europe during the 1960s, 1970s and 1980s favoured or encouraged movement out of town. This is certainly true of Thatcher's UK enterprise culture. Even where legislation did exist, such as France's Loi Royer (established in 1973), it was often difficult to stop the tide of movement out of town. Between 1973 and 1993, there were more than 7000 planning applications for large stores in France, more than one per day! As Metton (1995) notes, developers became accustomed to submitting more applications than they needed to compensate for refusals. They also became more skillful in understanding and manipulating the appeals process. Thus, since 1973, a further 750 hypermarkets opened in France (see also Cliquet 2000a). Guy (1998b) provides an excellent review of the tightening of recent planning laws across Europe (see also Poole *et al.* 2002a).

Such national planning legislation attempts to control the size and location of new developments. In addition, there is another aspect to the regulatory environment – legislation that governs the size of firms themselves. The nature of this type of legislation is also fundamental to the current power struggle in many countries of the world. In the United Kingdom, the retail battle described in the preceding text could not have taken place to the same degree in a very strict regulatory environment. The UK Competition Committee (formerly the Monopolies and Merger Commission) is not called into action unless a retailer manages to get towards 25% market share. Even major grocery retailers, such as Tesco and Sainsbury's, are still under this figure. Wrigley (1992, 1994) has contrasted this pliant UK environment with that of the United States in the same period. The much stricter Robinson–Patman Act and the Celler-Kefauver Act prevented US grocery firms from becoming as dominant in their markets as in many European countries. When these Acts were relaxed in the 1980s, we began to see the major battle for market dominance that has characterized US grocery retailing over the last 10 years. Although the US market is large and English speaking (partly accounting for the lure of European retailers: Hallsworth 1990), it is mainly the ending of this very harsh legislative environment that produced an attractive market. This came about, because the relaxing of the Robinson-Patman Act and the Celler-Kefauver Act produced an intense period of debt for many US grocers as they undertook leveraged buyouts and recapitalizations in order to strengthen their position in the market (i.e. they tried to make the firm stronger in an attempt to avoid hostile takeovers that would be more likely in the future in a less regulated environment). As Wrigley (1999) explains: 'A leveraged buy-out (LBO) occurs when third-party investors and/or managers of the firm offer to pay a premium over the prevailing market price of the firm and finance a change of corporate control by taking on a significant amount of debt' (p. 292). Between 1985 and 1988, 19 of the 50 largest US grocery retailers undertook LBOs, and corporate debt levels in the industry grew sharply (Wrigley 1999). With such debt burdens, many of the US grocery retailers could not expand geographically until cash reserves were built up again. Thus, this was the period when new entrants were most likely to do well in the US market and explains the most rapid period of growth for Sainsbury's and Ahold. It was not until the mid-1990s that the strongest players were free from debt to undertake the dramatic wave of consolidation that has characterized the US market in recent years. This intense period of (merger and acquisition) activity raised the market share of the top five firms by more than 50% in just three years (see Section 4.3 for an explicit discussion of mergers and acquisitions and their growing importance). For more details see Wrigley (1997,1999,2001).

Although out-of-town developments have taken place across Europe, planning legislation has been important in some locations, forcing new superstore development in existing centres or new centres, which offer a range of activities in addition to the superstore. These were especially common in areas identified by local planners as having low shopping provision rates. In France, these centres were called 'centres intercommuneaux' (see Reynolds 1993), and in the United Kingdom, district centres (see Guy 1994b).

The battle between retailers and planners is played locally as well as nationally. Local authorities have a duty to influence retail development in their localities in line with the kinds of national guidelines discussed in the preceding text. The extent of local power varies considerably across Europe. Guy (1998b) and Tewdwr-Jones (1997) argue that the United Kingdom is probably the most centralized regime (through legislation such as PPG6). Generally, local planners are concerned to monitor (cf. Guy 1996b):

total shopping needs (regulation of total retail floorspace)
the allocation of new retail floorspace
planning the requirement for new retail development
conditions for new retail development
protection and improvement of existing shopping centres
the development of retailing in comparison to the protection of the environment and other land use developments.

These are crucial requirements that mean local opportunities will always be available. That is, there will always be a local response to national planning guidelines, which reflects the existing retail landscape within that local environment. This helps to understand why certain developments are still granted planning permission, despite national legislation that seems anti-development. Finding these local opportunities will be very significant for major retail firms committed to growth in the future.

Changes in planning law can thus make severe impacts on retail dynamics. The revised PPG6 legislation in the United Kingdom has been a major factor in the slowing down of new superstore openings in the United Kingdom. This is examined in more detail in Section 3.3.

2.5 Conclusion

In this chapter, we have outlined some key trends that are helping shape the future retail environment. On the demand side, it is important that retailers keep abreast of the changing household dynamics. It is clear that generally consumers are more demanding. This has been dubbed the 'martini effect' – demanding goods any time, any place, and definitely at their convenience. At the same time, many consumers are becoming more discerning, and this has created much more niche markets that require to be targeted. The most successful retailers of the future will be able to tap into these niche markets far more effectively than they have done in the past. The geography of these changes will also be crucial. Some regions are experiencing population loss and a reduction in demand, while others are enjoying substantial wage inflation and economic success. In some regions the population is ageing rapidly. Retailers must understand the changing geography of consumer demand.

The supply-side changes are also revolutionizing the way in which the activity of shopping gets done. Consumers are faced with a smaller number of players in each market. Increasingly, these are global retailers bringing a similar range of products to customers all around the world. These firms are also crossing the boundaries between activities, becoming leading players in a wide variety of goods and services. As this level of competition hots up, retailers are becoming more price and cost conscious. At the same time, they are introducing new channels of distribution in order to satisfy the growing requirements of the customers more conveniently. Part of the challenge for the future is to manage all these channels in an optimal way, so that access to different types of customers in different types of locations (rural/urban, inner city/outer city etc.) can be maximized.

RETAIL LOCATIONS AND DISTRIBUTION CHANNELS: PAST, PRESENT AND FUTURE

3.1 Introduction

In this chapter, we wish to explore the major transformations in the retail landscape, partly brought about by the broad trends identified in Chapter 2. We also wish to look at the future of traditional retail locations, especially the high street, shopping centres or malls and retail parks. Although we have pointed out new planning legislations that prevent certain types of growth (Section 2.4), we argue that the traditional locations still have a lot to offer retailers and that much growth will take place in these environments (Sections 3.2–3.6). However, we also introduce new retail formats and distribution channels, and attempt to understand why geography or location is still crucial in understanding new distribution channels and why it forms a crucial part of the framework for successfully managing change and creating future retail landscapes (Sections 3.7–3.9).

3.2 The High Street or Town Centre

Despite the huge increase in out-of-town retailing (see next section) the high street still remains the retail heart of most towns and cities in Europe and will undoubtedly be a major retail landscape of the future. According to Field (1997), three quarters of all shopping is still done on the high street. That said, it is also true that the nature of retailing on the high street has changed considerably since the 1960s. An historical look at a range of European cities in the 1960s reveals a retail landscape dominated by clothing, footwear, jewellery, furniture and electrical goods. Food, fresh meat and vegetables were also prevalent. For example, an examination of the high streets of Leeds in the 1960s reveals the dominance of clothing, footwear and jewellery in the heart of the city centre, whereas away from the central area were increasingly larger numbers of wholesale food outlets, decorating products (DIY) and furniture stores. As more bulky goods such as DIY and electrical goods moved out of town (along with food), the nature of many high streets changed. It is now difficult to buy any of these goods on the high street. Today clothing and footwear is joined on the high street by a range of more niche outlets such as music shops, travel agencies, café bars and sports shops.

The buoyancy of fashion retailers, in particular, has helped maintain, and in many cases increase, the levels of rents and rates in many European high streets, and also make them popular destinations for other retailers. In the United Kingdom, for example, the growth of Next in the 1980s and 1990s almost parallels the growth of high street rents as they were prepared to pay very high premiums for the best pitches (which did, however, cause them some financial problems in the mid-1990s). Fernie *et al.* (1997) note how the recent influx of top fashion stores in the West End of London has shot rents to more than £250 per sq. ft. Figure 3.1 shows the current rents for many top European cities to be high and buoyant.

However, the future of the high street has long been a concern to planners, especially as fears of the competitive effects of out-of-town developments increased. In Europe, for example, there are fears that cities may follow US dynamics, where downtown cores are office and service centres as opposed to core retail areas. Therefore, there is now much interest in what makes a successful city centre, and how the city centre can remain an important future retail landscape. Peter Shearman Associates (1996) undertook an exercise to try and answer this question by comparing the cities of Hanover, Bordeaux and Bristol. They concluded that a successful city centre must:

1. create a broad retail offer to consumers
2. provide a thriving leisure environment
3. maintain large-scale office development
4. build high-quality public transport access

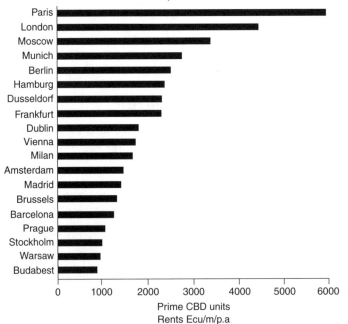

Figure 3.1 European retail rents. Source: Jones Lang Wootton, 2000

5. maintain accessibility for car owners
6. create compact, pedestrianized cities
7. restrict out-of-town developments

Thus in a sense, town and city centres need to reinvent themselves (Field 1997). The appointment of town centre managers in the United Kingdom will undoubtedly help implement such action plans in the future. It is clear that many city centres are fighting back. It is interesting to note that rents in Sheffield increased by 25% in 1997–1998, suggesting finally that the centre may be coming to terms with Meadowhall, the large regional out-of-town shopping centre only a few miles away.

The action plans identified by many city authorities are likely to indicate that the high street remains as a core retail landscape of the future. Opportunities will continue to appear as local planners revitalize older, more run-down areas of the city centre and build new arcades and waterfront developments. The latter is becoming a phenomenon in most modern world cities (see Goss 1996). Leeds has recently announced a £8 million redevelopment of Granary Wharf, while Manchester's canalside developments have transformed a run-down industrial environment into one of United Kingdom's most trendy bar and restaurant areas. In addition to the list of factors identified by the Shearman Associates in the preceding text, many would argue that residential (re)development is also crucially important. It is clear that city centres outside the United Kingdom have had a head start in this respect. Many cities such as Hanover and Bordeaux have always had a significant residential population, often living in flats above the main shopping arteries. These types of development are laying the foundations for concepts such as the 24-hour city. The combination of office development, residential development and more leisure facilities means that people can enjoy the attractions of the city at all hours of the day.

It is likely that the high street will continue to see new fascias and perhaps more stores of traditional fascias. Many traditional and new high street retailers in the United Kingdom have announced ambitious growth plans. In 1999, Debenhams announced plans to open a further 15 stores over the next four years. Oasis stores is to open nine new fashion stores to take its high street total to 140. Clinton Cards is planning to open 40 new greeting card stores a year for five years, and Whitbread is planning both 50 new Costa coffee shops and 100 new Pizza Hut stores by 2001. The New Covent Garden Soup Company is planning to open 200 new soup bars on high streets throughout the United Kingdom by 2004. In 2001, the following organizations announced major expansions: Merchant Retail (from 67 stores to 150), Officers Club (150 new stores over 3 years), Greggs (from 1100 to 1700), Fast retailing (150 new stores), Hair Cuttery (100 new stores) and Coffee Republic (from 76 to 200 by 2004).

The influx of high fashion retailers is also beginning to boost other retail centres outside London. Diesel, Versace, Tommy Hilfiger and Calvin Klein are rumoured to be interested in King Street in Manchester, following the letting to DKNY in the late 1990s. Other European fashion retailers such as Mango, Zara, River Island, Hennes and Mauritz are all said to be seeking large amounts of UK floorspace (Cavanagh 1999). There are likely to be opportunities for other European countries to attract major fashion retailers in the future. Lamey (1997) notes that Germany is now attracting European players in the boutique sector, mainly as a result of the new wave of luxury shopping malls (especially CentreO near Dusseldorf). She cites the example of New Look opening its first outlet in Germany at CentreO, but announcing at the same time that they estimate Germany could support 500 of their stores. Oasis too is reported to be looking at 60 to 70 stores in Germany over the next ten years. (Lamey 1997, p.16).

However, many analysts predict that some retail activities face difficult times ahead and may well disappear from the high street. According to Cope (1996), Internet/telephone shopping will, by 2006, cause the demise of the high street banks, travel agencies, video shops and record outlets. Other variety type stores may also disappear as they too face competition from Internet shopping. Littlewoods, for example, put all of their stores up for sale in March 1997. Field (1997) responds to these issues by speculating on the possible replacement stores of the future: wine bars, restaurants, hair salons, pharmacists, delicatessens, fishmongers, cake shops, high fashion boutiques, burger bars and food convenience stores. He believes that these activities could also revive the smaller high streets associated with small and medium sized towns.

Part of this process of transformation will be achieved by large, international retailers, and part by independents. Many of the more specialist shopping areas of our major cities now encourage small, independent retailers with niche products. The condom shop in the Corn Exchange at Leeds is a good illustration. Field (1997) talks about the use of these types of outlets to stimulate retailing by appealing to very different sorts of community. Such specialist shopping areas may well help reduce the existing differences between primary areas and secondary areas. Primary areas are likely to remain the strongholds of major multiple retailers. These more specialist shops could be a way of revitalizing secondary areas, which are often currently occupied by cut-price stores, charity shops and market stalls.

3.3   The Superstore and Hypermarket

The conditions described in Sections 2.3.1 and 2.4 allowed an unprecedented period for the growth of superstores in the late 1970s and 1980s in many developed countries. Figure 3.2 shows the growth over the key period 1975 to 1995 in a variety of European countries.

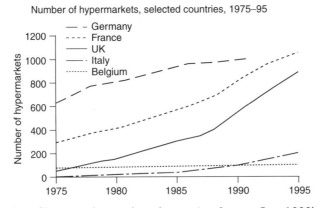

Figure 3.2   Number of hypermarkets, selected countries. Source: Guy, 1998b

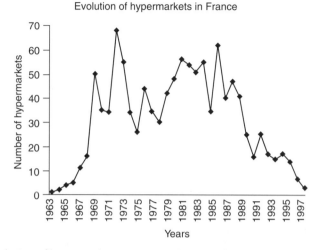

Figure 3.3 Evolution of hypermarkets in France. Source: Cliquet (2000a)

However, it is clear that hypermarket growth has slowed down significantly since the late 1980s across Europe. Figure 3.3 shows the dramatic decline of new developments in France. Much more stringent legislation in 1996 (Loi Raffarin) has finally it seems stopped major new store developments. Similarly, the 'Ley de Ordenacion del Comercio' in 1995 has made it much harder to obtain planning permission for out-of-town developments in Spain. The strictness of these new legislative environments will have a major impact on future retail landscapes. Besides preventing further growth of French companies, foreign retailers too will be discouraged from entering the French market. Even in countries such as the former East Germany (which allowed a period of laissez faire activity after reunification) new out-of-town developments have been banned. The same is true in Portugal and Belgium (which insists that an impact study must be carried out by the applicant). In the United Kingdom, the revised Planning and Policy Guideline (PPG) 6 legislation has also tightened up on out-of-town developments. European laws are now beginning to harmonize and agree that unrestricted retail development is harmful to existing retail infrastructure. This means that sites with existing planning permission are especially attractive and expensive (see the discussion on retail warehouses in the following text). Guy (1998b) also notes the recent trend of reducing the upper size limit of new stores, especially in Belgium, France, Portugal, and Spain. The French, for example, have recently set an upper limit of only 300 sq. m.

However, in some countries there are signs that there may well be a more liberal planning regime in the years to come. Italy provides a good example. It has often been said that Italy has been Western Europe's most regulated country. In 1998, the new left-wing government presented a plan to throw the business of retailing open to the market. This means that no permission is now required to begin small store trading, there are longer opening hours, and a simpler food, non-food retail category has been introduced (replacing 14 categories of retailing that required a licence for each). However, the plan is unlikely to extend to more freedom for the large store format. Indeed, a new law was passed in April 1998 to re-evaluate the opening of new planned shopping centres. This is good news for the incumbent Italian retailers who understand they are small by other European standards, and are continually requesting time to grow in their home market before having to face foreign competitors (Pellegrini 1995).

However, it is evident that despite the tighter planning legislation referred to above, new store developments are continuing. A search through recent UK press releases and Web sites reveals much expansion. In 2000, Tesco alone opened 3 new 'Extra' stores at Wrexham, Leeds and Cambridge, and announced plans for another large store at Walsall. Morrisons opened 12 new superstores in 2000 and plan to open a further 7 in 2001 and 4 in 2002. Asda has opened 60 new superstores between 1996 and 2001, with plans for another 30 to 40 over the next three years. Outside the United Kingdom, Lidl is poised to open 25 new supermarkets in the supposedly saturated market of the Netherlands.

The question is how this can happen, given the tighter legislation discussed in Section 2.4? A number of explanations are plausible. First, retailers could have acquired the land prior to the revised legislations becoming operational. Thus, they already possess planning permission on that land. Second, it seems clear that the power in planning enquiries and, indeed, in the drafting of planning law does not simply lie with the local or national government. Hallsworth (1997) and Pal *et al.* (2001) suggest that retailers have been very proactive in helping shape legislation. In particular, Pal *et al.* (2001) argue that they have encouraged a sense of vagueness in the planning documents, a vagueness that allows them to exploit loopholes when negotiating with the Government at planning appeals. Third, it may well be the case that local authorities actually wish to promote retail development in their localities, despite national planning guidelines that discourage this. Retailer planners across Europe are still faced with the issue of trying to equalize accessibility to retail facilities at the local level. In addition, there are usually good tax reasons why local authorities may wish to encourage development within their localities.

Metton (1995) notes that the power of the private sector to choose the most desirable locations has led to 'commercial wastelands'. The late 1990s has also seen huge interest in the United Kingdom in so-called *food deserts*, areas where many lower income households (including low mobility households such as pensioners and the disabled: areas that the private sector has generally shied away from) face poor accessibility to good quality grocery retailing. It is likely that planners will continue to look for retailers who would be prepared to invest facilities in these locations (perhaps for reduced rent or land costs). Metton (1995 p74) described the initiatives to revive rural commerce in many of the commercial wastelands in France and concludes that the future implementation of Loi Royer is likely to be very different in different parts of France. Similarly, local planners in Leeds have recently given planning permission for a large 90 000 sq.ft Tesco store in Seacroft, on the outskirts of the City. Permission was granted solely because it is deemed to be in a 'food desert' area (and it is part of a wider scheme by Tesco to invest in inner city areas: the 'London Initiative', for example, entails a 20 million pound refurbishment scheme in Beckton, involving Tesco and a range of public and private sector organizations). It will be interesting to see how such a superstore trades in such a low-income environment (see more discussion on food deserts in the following text).

Despite the success of superstore operators in continuing to open stores, it is clear that the rate of growth has declined. Along with the stricter planning regimes outlined in the preceding text, superstore operators have faced stiffer competition from discounters (see Section 3.4), and in some countries, have experienced problems associated with recession and overvaluation of stock and assets (see Wrigley 1996). Poole *et al.* (2002a) explore the extent to which these conditions have varied across Europe. How have the superstore operators responded to these difficulties?

One major response has been to increase investment in refurbishment and store enlargements. In the 1990s, Tesco announced plans to refurbish a large number of their now older superstores. They will add significant new floorspace to their asset base. In some cases, significant extensions will allow stores to be revamped according to a new 'Tesco Plus' fascia (with significantly extended product ranges). The first opened in Essex in 1998, and at 90 000 sq.ft, is currently the largest in the United Kingdom. Perhaps the biggest refurbishment programme in recent years has come from Safeways. In 2001 they announced a 450 million pound refurbishment programme for 100 of their stores.

A second major response has been to diversify into non-traditional supermarket product lines. The modern superstore has seen a fresh injection of new products and longer opening hours. Tesco now operate financial services and their share of the petrol market is growing fast (Bennett 1998). The product range is growing continuously. Plans for stores to stock motor scooters and Intel Pentium computers in 1999 are now in operation. Also, by 1999, 100 of their 568 stores have greatly extended opening hours (including 81 stores that now operate 24 hours per day). In addition, they operate a rigorous price campaign called *unbeatable value* and a very successful loyalty card called *Clubcard*. Asda too is increasingly devoting more of its floorspace to non-food product lines, especially since the takeover by Wal-Mart (Burt and Sparks 2001). Items such as clothing, brown and white goods, books, CDs, and so on are now commonly available in large stores. Of course, this has been the case for a long time in some European hypermarket formats, especially in Belgium and France. The activities of GIB in Belgium now include groceries, DIY, toys, perfumes, stationery, clothing and fast food. ASDA plan to increase non-food activities through their new 'Supercentre' concept, which will increase the size of existing stores to between 100 000 and 130 000 sq.ft, and offer computer software and hardware, small electrical goods, branded sportswear and sports goods for the first time (Mills 1999). Perhaps, more interesting has been the development of what might be seen as rather radical new products and service offerings, particularly in the financial services sector. Both Tesco and Sainsburys in the United Kingdom are now offering a limited range of banking products in their stores, which, although branded as 'Sainsbury's Bank and Tesco Bank' are, in fact, provided by major financial services providers such as the Royal Bank of Scotland (also see Alexander and Pollard 2000).

A third response has been to invent new formats. This takes us away from discussion on superstores and will be looked at in Section 3.8.

The key question is where might future stores be built that will gain planning permission? The answer is almost certainly tied up with the food desert argument raised in the preceding text. This is worth elaborating on further. As we noted in the preceding text, there is concern the world over about the impacts of the superstore on traditional town and city centres. In addition, there is increasing concern that the superstore developers have neglected certain parts of towns and cities as they search for sites containing more affluent consumers. It seems to be the case that in many towns and cities, the geography of superstore development has not been spatially uniform and access to such stores is highly variable. This has led to a new term in the literature–the food desert. Such areas may offer the best opportunities for retailers to build larger stores in the future, as this is likely to be encouraged by local planners. But how can these be measured or recognized?

43

One such study to measure and identify food deserts is provided by Clarke *et al.* (2002). They used performance indicators derived from a spatial interaction model (see Chapter 8). Figure 3.4 shows the 'effectiveness of delivery' of grocery retailing to residential locations. This performance indicator was introduced by Clarke and Wilson (1994). The higher the indicator score, the greater the degree of accessibility. Figure 3.4 maps this indicator for the Leeds/Bradford area. The map clearly shows pocket of poor accessibility that we could now label as food deserts (though see Clarke *et al.* 2002 for a more thorough discussion). Eight such areas are identified in Leeds and Bradford. These areas tend to be urban concentrations located towards the edge of the conurbations, which therefore have poor access to city centre food retailing opportunities, and are also located away from out-of-town superstores. It should be recognized that these are areas of poor provision relative to the rest of the city. However, simply having relatively poor access to grocery provision is unlikely to convince planners that all of these areas are food deserts. An area might only be classified as a food desert if the residents of that area have little or no means of travelling significant distances to purchase food (see Williams and Hubbard 2001 and Piacentini *et al.* 2001 for accounts of how residents in deprived estates have very poor accessibility levels). It is therefore necessary to look at the social class structure of the

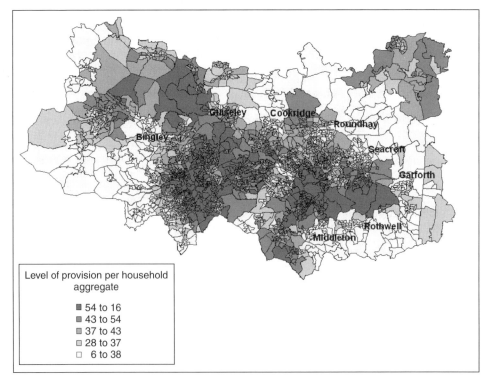

Figure 3.4 Level of provision per household in Leeds/Bradford. Source: Clarke *et al.* (2002)

Table 3.1 Social composition of possible food desserts in Leeds and Bradford

	%AB	%DE	% Retired	% Inactive	% No car
Leeds/Bradford	21.6	11.9	28.0	12.4	39.7
Bingley	31.5	7.9	28.8	10.4	32.5
Cookridge	26.8	6.3	32.7	11.6	30.9
Garforth	20.5	11.7	27.1	7.3	26.2
Guiseley	27.5	13.0	28.3	5.8	35.2
Middleton	6.8	20.6	24.3	15.8	61.2
Rothwell	16.0	11.6	32.9	10.2	38.1
Roudhay	30.6	6.6	30.1	13.5	26.2
Seacroft	5.6	15.4	32.8	17.4	60.9

Source: Clarke *et al.* (2002)

households in each of the areas of poor provision in order to identify food deserts. Areas with a high percentage of DE, retired and inactive households will be less likely to own a car and therefore more reliant on grocery shops in their local area.

Table 3.1 shows the social composition of the areas of poor provision in Leeds and Bradford, as well as the average composition of the area. Of the eight areas of poor provision, two (Middleton and Seacroft) stand out as containing a low proportion of AB households and a percentage of DE and inactive households well above the average for the Leeds/Bradford area. These are also areas where more people do not have access to a car and are therefore reliant on stores in their local area. The poor level of provision in these areas then might constitute a food desert. The other areas in Leeds and Bradford that experience poor provision contain a higher proportion of AB households and a lower percentage of households with no car. Residents of these areas are therefore more likely to be able to travel to purchase food and are less reliant on the poor facilities in their local area. Perhaps these areas should not be identified as problematic food deserts, although for some residents the problems are as great as those we find in Middleton and Seacroft.

The same concerns over food deserts exist outside UK towns and cities. Nayga and Weinberg (1999) provide a good US illustration. They observe:

> One of the most pressing problems in Urban America today is the serious shortage of supermarkets in many inner cities despite America's abundant food supply and state-of-the-art food distribution system. (p141).

They go on to explain the consequences of this problem. First, such inner city environments are left only with independent stores with their much higher prices. Second, although discount retailers may find a niche market in such areas (see Section 3.4), there is concern that such retailers provide less fresh produce and less choice of produce, thus resulting in poorer diets. This issue is explored in relation to the Leeds case study outlined above by Wrigley *et al.* (2002).

3.4 The Discount Store

In the previous section, we argued that superstore retailers have tended to search for green-field sites in and around the more affluent suburbs of western towns and cities. As Wrigley and Clarke (1999) argue, those strategies left two interrelated gaps in the marketplace. First, a 'value platform' gap–a vulnerability to low margin, limited-line discounters. Second, a 'locational' gap–urban high streets and less affluent suburban areas felt to be too unprofitable for traditional superstore operations (until more recently–see preceding text). In most western countries, some of these gaps were filled by the arrival of the discount store. The importance of the discount store in terms of total sales varies between countries. Figure 3.5 shows the penetration of deep (or hard) discounters across Europe in the late 1990s. As the graph shows, Germany, Austria and Belgium lead the way, with Germany having some of the most important discount retailers in Europe.

To understand the geography of discount retailing, it is useful to look at the United Kingdom as an example. As the major superstore retailers grew in the 1980s, only a few retailers chose to exploit the value platform and locational gaps mentioned in the preceding text. In England and Wales, Lo-Cost and Kwik Save emerged as important players in towns and cities across Wales, central and northern England in particular, while Shoprite began to make inroads into market share in urban areas of Scotland. The growth of Kwik Save and Shoprite is discussed in detail by Sparks (1990, 1996b respectively). The success of these companies was significant as they enjoyed almost monopoly positions in the poorest areas of the United Kingdom (facing competition from independent retailers and local Co-ops rather than the major superstore players). Then, in the early 1990s, the United Kingdom saw the invasion of more discount retailers from continental-Europe, led by Aldi and Netto (Burt and Sparks 1994, 1995), and followed by Lidl and Ed. As Figure 3.5 shows, the discount retailers were already well established in other parts of Europe. They were attracted to the United Kingdom because of the success of the incumbent UK discount operators and the relatively high profit margins available in UK grocery retailing at the time (typically 6% margins compared with 2% in many other European countries). Not surprisingly, Netto and Aldi targetted the very areas where the UK discounters were present–the heart of less affluent urban areas. Figure 3.6 shows the growth and location of Aldi stores during the 1990s. From their initial strongholds of Birmingham and Manchester, they became a significant retailer in most parts of northern and central England by 2000.

The impact of these deep discounters was significant on all grocery retailers in the United Kingdom. First, they had an immediate impact on Kwik Save and Shoprite, as they

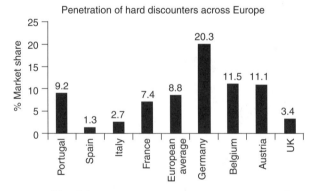

Figure 3.5 Penetration of hard discounters across Europe. Source: Wrigley and Clarke (1999)

competed head on in the traditional heartlands of the UK retailers. Shoprite were in serious trouble by 1994 (Sparks 1996b), and, in 1998, Kwik Save was effectively taken over by Somerfield. As Wrigley and Clarke (1999 p1) maintain, 'This was a logical outcome of an increasingly desperate struggle by Kwik Save to arrest an accelerating decline in its like-for-like sales, operating profits and market capitalization.' Second, even the large superstore retailers began to feel the heat, forcing them to review their pricing structures. However, the outcome of this review produced interesting geographical variations in the response of the major players. Prices were often reduced in areas where deep discounters were present (but not in other areas where the retailers enjoyed monopoly or duopoly power–a fact that partly led to the Monopolies Commission report on the grocery market in 1999, which expressed considerable concern over monopoly retail power in many areas of the United Kingdom). In the poorest urban areas, where the battle was fiercest, the major players even reconfigured to limited-line or 'discount superstore' fascias. For example, Asda announced

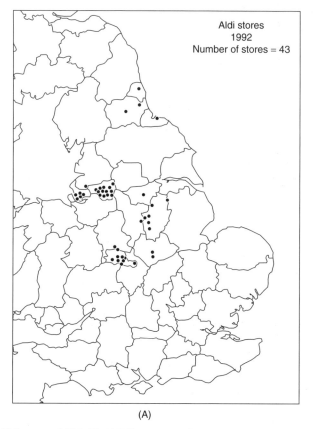

(A)

Figure 3.6 (A) Aldi stores 1992 (B) Aldi floorspace share in 2000. Source: Wrigley and Clarke (1999) & Poole *et al.* (2002a)

47

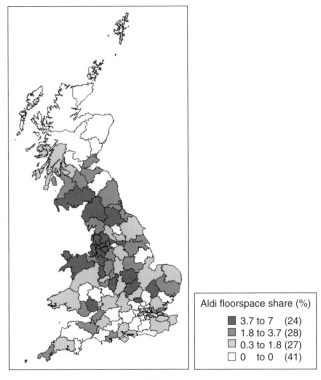

Aldi floorspace share (%)

■ 3.7 to 7 (24)
■ 1.8 to 3.7 (28)
□ 0.3 to 1.8 (27)
□ 0 to 0 (41)

(B)

Figure 3.6 (*continued*)

the conversion of many northern stores (closest to inner city areas) to the Dales Discount format. By late 1993/1994 then, it was clear that the entry of the limited-line discounters had help trigger a profoundly destabilizing shift in competitive conditions in UK food retailing (Wrigley 1994). However, the competitive response of the major players (mainly the concentration on price for many day-to-day items) meant that the share of the deep discount sector peaked in the United Kingdom in 1994 at around 10.0%. As Table 3.5 shows, by 1997/98, the discounters' share of the packaged grocery market had fallen significantly, although by the end of 1998 they had opened 438 stores between them.

However, it is interesting to note that the penetration of the discounters in the south of England by 1998 was still very low. It is an interesting geographical question to ask where the discount market could be exploited further over the next ten years, especially given the fact that many of the Kwik Save stores taken over by Somerfield are likely to be targeted for disposal. So the key question to be briefly looked at here is do market opportunities still exist for limited-line food discounter expansion? The analysis that follows is based on Wrigley and Clarke (1999). Their technique for identifying potential sites begins by establishing that in 1996 the average number of socio-economic group D and E households (the lowest socio-economic categories in the UK census) in postal districts that contained a limited–line discount store lay in the range 1500 to 2000. They then established a 'potential demand' map, defined on the basis of the number of postal sectors that have 500 or more D and E households within 5 minutes drive time of their centroids (Figure 3.7).

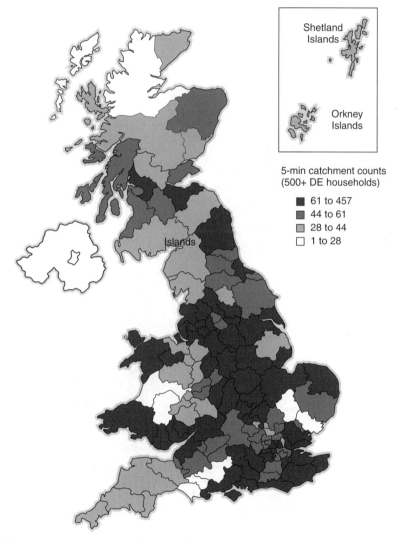

Figure 3.7   Catchment counts–DE households. Source: Wrigley and Clarke (1999)

Clearly, some areas identified in Figure 3.7 will already have a discount retailer present in the catchment area. Figure 3.8 thus maps the areas of unserved demand, which therefore offer the greatest opportunities to limited-line discounters. Those areas are most heavily represented in southern England, including East Anglia and parts of the southwest.

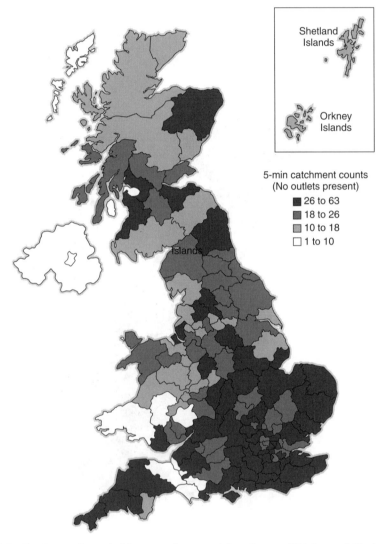

Figure 3.8 Catchment household counts–low provision. Source: Wrigley and Clarke (1999)

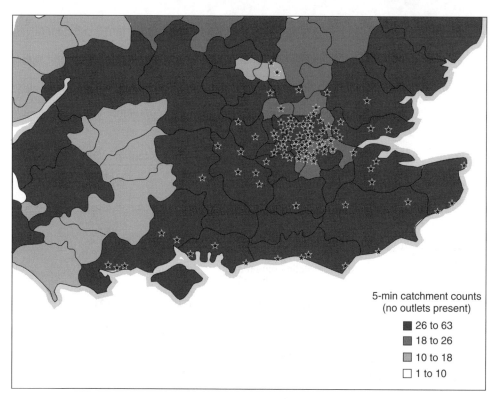

Figure 3.9 Catchment counts (detail). Source: Wrigley and Clarke (1999)

Finally, Figure 3.9 focuses on the southeast and, hence, prioritizes the previous map of opportunity (Figure 3.8). Figure 3.9 identifies the best 100 target sites (out of more than 1000 possibilities) for a new limited-line discounter. 'It is our view that the future scale and direction of the discount food retail sector rests to a large extent on how real the potential of such sites turns out to be' (Wrigley and Clarke 1999 p14).

3.5 The Shopping Centre or Mall

The shopping mall is another US invention that began to grow in Europe from the 1960s onwards. Today the shopping malls of Europe tend to have the highest turnovers of any of the retail formats. According to the UK Index of Retail Trading Locations published in 1998 by DTZ, the malls at Meadowhall (Sheffield), Merry Hill (Birmingham), Metro Centre (Gateshead) and Lakeside (Essex) are the most successful retail locations in the United

Kingdom in terms of turnover. The analogy has been made between these centres and what might be termed *edge cities*, such is their size and importance in the retail hierarchy (see Lowe 2000).

According to Guy (1998b), around 75 of the 108 shopping centres in Europe larger than 50 000 sq. m lie in France, United Kingdom and Germany. He identifies two key stages of development: 1970 to 1975 and 1990 to 1995. Table 3.2 lists the location of the major shopping centre developments in Europe. The vast majority are off-centre or out of town, although there are many examples of city centre developments. Some have

Table 3.2 The Super League[a] of European shopping centres, 1997

Country	Centre	Space (sq m)
UK	Metro Centre (Gateshead)	140 000
France	Grand Littoral (Marseille)	140 000
France	Creteil Soleil (Paris)	140 000
Portugal	Colombo (Lisbon)	133 000
UK	Merry Hill (Dudley)	130 000
France	Belle Epine (Paris)	130 000
UK	Meadowhall (Sheffield)	116 000
Austria	Shopping City Sud (Vosendorf)	115 000
Spain	Nuevo Centro (Valencia)	112 000
Germany	CentrO (Oberhausen)	100 000
UK	Lakeside (Thurrock)	93 000
Germany	Saale-Park (Leipzig)	93 000
Ireland	Blanchardstown (Dublin)	90 000
Spain	ABC Serrano (Madrid)	90 000
Italy	Shopville (Turin)	82 000
Germany	Flora-Park (Magdeburg)	80 000
UK	Brent Cross (London)	79 000
The Netherlands	In De Boogard (Rijswijk)	70 000
Turkey	Icerenkoy (Istanbul)	68 000
The Netherlands	Hoog Catharijne (Utrecht)	65 000
Norway	Stovnersenter (Oslo)	60 000
Denmark	City 2 (Copenhagen)	58 000
Italy	Auchan (Milan)	57 600
Italy	Valecenter (Venice)	56 000
Sweden	Frolunda Torg (Vastra Frolunda)	54 000
Belgium	City 2 (Brussels)	53 600
Spain	Glories (Madrid)	53 000
The Netherlands	Zuidplein (Rotterdam)	52 000
Hungary	Polus Ring (Budapest)	52 000
Switzerland	Glatt-Zentrum (Zurich)	45 300
Finland	Itakeskus (Helsinki)	36 000
Czech Republic	Praha West (Prague)	32 000
Greece	Continent (Thessalonika)	22 000
Poland	Globe Trade Center (Warsaw)	na

[a]See text for explanation.
Source: Corporate intelligence on Retailing, 1997.

also been built to serve new settlements such as Creteil in France and Milton Keynes in the United Kingdom (Guy 1998b p 963). The latest mall to open in the United Kingdom is Bluewater Park in Kent. This development includes 320 shops, three leisure villages with cafes and restaurants, an entertainment centre and a 12-screen cinema complex. The entire site is surrounded by 50 acres of landscaped parkland. In Germany, CentreO now vies with Bluewater for the distinction of the largest centre in Europe. Built on the site of a former steel mill near Dusseldorf, the retail centre provides parking for 11 000 cars. International retailers include Gap, Coca-Cola, Levis, Warners, Planet Holly-wood and Ronnie Scotts. The centre is as much a theme park, as it is a retail mall (see also Chapter 3). Other recent major developments include the Colombo Cente in Lisbon and the Grand Littoral centre in Marseilles. The Diagnol Mar in Barcelona was scheduled to open in 2000.

A key question is where may development take place in the future? It is clear that many European countries still have economies dominated by small shops. Figure 3.10 shows the number of retail shopping centres by country in Europe and the number of retail outlets per thousand habitants. The table shows that Greece, Portugal, Spain, Belgium and the Netherlands have the highest total number of centres. There is currently 69 m sq.ft of lettable floorspace in Europe, with France and the United Kingdom ahead of the rest. On a per capita basis, Sweden, the Netherlands and France lead the way. These variations are more prominent within countries. Table 3.3 shows the variations in the distribution of shopping centres in Austria in the mid-1990s. If profits can be made in Vienna, Lower Austria and Tyrol (where there are high floorspace rates per resident), then surely there are opportunities in Burgenland, Carinthia and Vorarlberg, even if these have relatively less income per capita.

However, given the tighter planning legislation across Europe, it is likely that the pace of development will slow down rapidly. Indeed, in some countries there is concern that there may now be too many shopping centres, which will inevitably lead to closures. Borchert (1995) discusses a recent stocktaking study in the Netherlands that concluded that there was no future for approximately half of the existing neighbourhood shopping centres. However, it seems, that in the short term at least, there are many new developments in

Table 3.3 Distribution of shopping centres in Austria

Land	Number	Selling area (m²)	Annual turnover (billion Schillings)	m²/100 inhabitants
Vienna	23	262 000	12.0	17
Lower Austria	17	267 800	11.4	18
Burgenland	3	13 700	0.6	5
Upper Austria	16	170 000	6.7	13
Salzburg	8	72 600	2.6	15
Tyrol	9	104 900	4.2	17
Vorarlberg	4	33 500	2.0	10
Styria	14	151 200	5.4	16
Carinthia	6	44 000	1.9	8
Total	100	1 119 700	46.8	

Source: Nielsen, quoted in *Linzer Volksblatt*, 20 June 1994: 22.

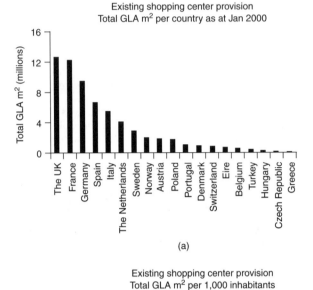

Figure 3.10 Shopping centre provision. Source: Hillier Parker

the pipeline across Europe as a whole. Figure 3.11 shows European variations in shopping centre developments in the pipeline in mid-1998. There was an estimated 8.2 m sq.ft of floorspace under development. The United Kingdom led the field with 1.4 m sq.ft in the pipeline, closely followed by Italy and Germany, with 1.2 m sq.ft each. Despite the pessimistic noises coming from the Netherlands (see proceeding text), there was still new opportunities for growth, although these developments are largely planned for existing

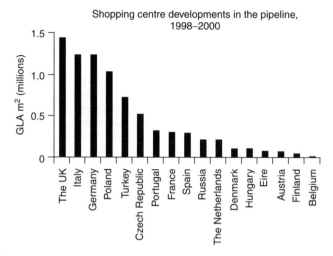

Shopping centre developments in the pipeline, 1998–2000

Figure 3.11 Planned development of shopping centres. Source: Hillier Parker

town or city centres. Many southern and eastern European countries had a relatively high proportion of shopping centre development owing to their less restrictive planning regimes and interest shown in these areas by foreign investors. A useful summary of the opportunities offered by Central Europe is provided by Jones Lang Wootton and Oxford Institute of Retail Management (OXIRM) (1998).

It is likely that floorspace will also continue to grow through redevelopment program- mes. In the United Kingdom, in the mid-1999s, there were a number of major redevelop- ment programmes under construction, including Serpentine Green in Peterborough, Prince Bishop Centre in Durham, Priory Meadow in Hastings, Oracle Centre in Reading, Birming- ham Bull Ring, Thistle markets in Stirling and Buchanan Galleries in Glasgow. The National

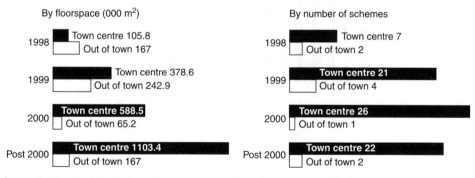

Figure 3.12 Predicted shopping centre openings. Source: Hillier Parker

Retail Planning Forum (1998) discusses a number of the major UK schemes in the pipeline, and Figure 3.12 summarizes the predicted growth in floorspace in the United Kingdom alone.

3.6 The Retail Warehouse and Retail Park

As with all forms of retail format, there are widespread variations in the provision of retail warehousing. Retail warehouses tend to be single developments, while retail parks are clusters of individual retail warehouses. Figure 3.13 shows how Belgium, France and the United Kingdom top the European league table. Guy (1998a) identifies a number of waves of retail warehouse developments in Europe.

1. Mid-1970s onwards–household and DIY
2. Early 1990s–electrical retailers
3. 1995/96–more traditional high street retailers (Marks and Spencers, JJB Sports, Shoe City, etc.) plus new leisure activities and a new wave of bulky goods, such as pet equipment and office supplies

Table 3.4 summarizes the range of retail sectors now investing in retail warehouses and the variations in these across Europe. It is likely that the gaps in this matrix will be filled as we enter the next Millennium.

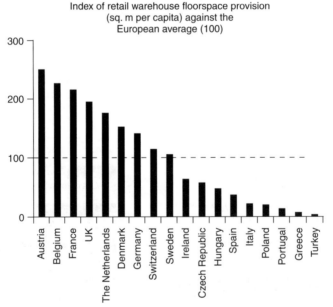

Figure 3.13 Index of retail warehouse provision. Source: Healey and Baker

Table 3.4 Table showing retail warehouse occupiers in Europe, by sectors

Country	DIY	Furniture/ carpets	Electrical goods	Car accessories	Clothing	Footwear	Sports goods	Other	Typical anchor
Austria	●	●	●	●				●	Hypermarket DIY furniture
Belgium	●	●	●	●	●	●	●	●	Hypermarket DIY
Czech Republic	●	●	●	●	●	●	●		Hypermarket DIY
Denmark	●	●	●		●			●	Hypermarket furniture
France	●	●	●	●		●	●	●	Hypermarket DIY electrical
Germany	●	●	●	●	●	●		●	Hypermarket DIY furniture
Hungary	●	●	●					●	Hypermarket DIY electrical
Italy	●	●	●	●	●	●		●	Hypermarket furniture
The Netherlands	●	●	●	●				●	DIY Furniture electrical
Poland	●	●	●	●				●	Hypermarket furniture
Portugal	●	●	●	●				●	Hypermarket DIY
Spain	●	●	●	●	●		●	●	Hypermarket furniture
Sweden	●	●	●	●	●	●	●	●	Hypermarket furniture
Switzerland	●	●	●					●	Hypermarket
UK	●	●	●	●	●	●	●	●	Supermarket DIY Some without anchors

Source: Healey and Baker

57

The future of the retail park looks very secure. In the United Kingdom, B&Q is planning to open 25 new warehouse stores in 2001. JJB Sports announced their plans in 2001 to expand the chain of superstores from 214 to 500 over the next five years. Similarly, MFI are planning 60 new outlets over the next three years and Matalan plan to grow from 123 to 240 outlets over the next eight years. With such growth, the retail park has become more sophisticated in recent years. Guy (2000) talks about the transition from 'crinkly sheds to fashion parks', which are extremely attractive to investors. Rental growth is very buoyant in most European countries and reached record levels in the United Kingdom in 1996. For example, between 1990 and 1995, UK high street prime rentals fell by 9% on average, whereas retail warehouse rents grew by 34%. The reason for this growth is that demand is high, whereas supply is limited. Demand is high because as Table 3.4 suggests, there is a new wave of products that are moving out of town: pet stores, computer outlets, toys and so on. In addition, newer retail warehouses require less floorspace (10 000 to 15 000 sq.ft as opposed to 25 000 to 40 000 sq.ft). This has led to subdivision of original premises in many older retail parks and allowed the new demand to be satisfied, thereby increasing rents. Lawson (1996) predicts a new wave of redevelopments mainly consisting of subdivisions into many smaller retail warehouses. The most promising sites for the future in the United Kingdom are those that have 'open A1 consent'. These are parks that are not restricted to particular goods only. They are generally parks developed before the publication of the revised PPG6 legislation and now enjoy three to four times the level of rents typical of restricted retail parks. The new legislation will not generally allow 'open A1 consent'. Thus, as Guy (1998a) notes, 'Occupiers seeking to expand into new geographical areas may as a result have their last chance for several years to find the spacious and accessible premises they require (p.10).'

What about prospects in other European countries? Healy and Baker (1998) note that in Italy, the concept of the retail park is barely established and investment activity is limited. This is also true in Spain and Portugal, although both countries have witnessed new interest in retail warehouse development. The best growth prospects would seem to be in central and eastern Europe, especially Poland, Hungary, the Czech republic and Turkey. These areas are forecast to be major growth areas for the sale of white electrical and DIY goods. Already, firms such as Praktiker, Castorama and Obi have gained footholds in these markets. Praktiker are also opening new retail warehouses in Greece and Turkey.

In an interesting new trend in the United States, a small number of very large retail warehouse operations are clustering together to form so-called *power centres* (Bodkin and Lord 1997, Hahn 2000, Jones and Doucat 2000). Interestingly, research suggests these are not strongly integrated centres in the traditional sense. According to Bodkin and Lord (1997), there is little cross shopping at these centres and, hence, they are best described only as a 'loosely connected' group of stores.

3.7 New Retail Locations

There is a growing interest in the retail geography literature on new retail spaces or locations. The first new space of interest has been centres of communication and transport, especially airports and railway stations. Freathy and O'Connell (1998) describe the growth in airport retailing and how this area might develop in the future. Interestingly, Connex, the UK transport company, announced a 15 million pound package to introduce convenience stores in major railway stations in the South of England in 2001.

Another major new development has been the factory shop. Indeed, Norris (1999) states that since the 1980s, factory outlet shopping has been the most dynamic growth sector in the European property market. Figure 3.14 shows the rapid growth across Europe since 1990. Factory outlets are generally discount superstores selling a wide range of non-food goods. As the name suggests, most are manufacturer outlets. This helps manufacturers sell direct to the public, normally through the sale of surplus or out-of-date merchandize. Locations for factory outlets are usually chosen so as not to compete with 'up-market' or major high street locations–hence the location of such outlets in places such as Hornsea and Street in the United Kingdom. Despite the new stricter planning regime, many analysts believe they will still grow substantially across Europe in the future. Figure 3.15 shows the variations in existing and planned developments of factory outlets across Europe in 2000. A new company called Festival Parks Europe is being set up to build and operate 10 large factory shopping and entertainment centres across Europe. The first was opened in Majorca in 2000.

As Crewe (2000) points out, more recently, a range of unconventional spaces and practices have been recast as legitimate areas of study in retail geography. Of particular interest here has been her own work with Gregson (Crewe and Gregson 1998, Gregson and Crewe 1997) on the car boot sale. Crewe and Gregson (1998) explain that the car boot sale (with its haggling, fun, and unpredictability) provides a 'compelling alternative to the uniformity and predictability of the mall' (p.50). Such spaces are important as they bring a sense of 'quasi-fun and quasi-leisure', and hence support the belief that for many consumers shopping can be social and recreational rather than simply economic and passive. In many ways, therefore, such new spaces resonate more with the excitement and unpredictability of traditional markets and fairs than they do with modern superstores and malls. There is an interesting geography to explore here–do the locations of these matter in a large urban area? In other words, which type of consumer may be most attracted to these sales and where are they located?

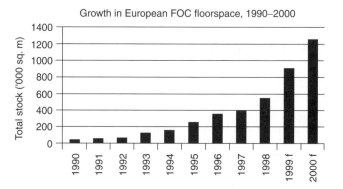

Figure 3.14 Growth in European Factory Outlet Centres (FOC) floorspace. Source: DTZ Debenham Thorpe (1999) Research Department

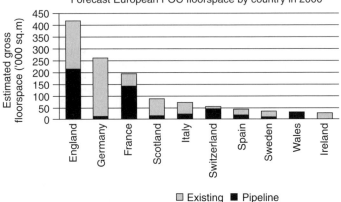

Figure 3.15 Forecasts of European FOC floorspace. Source: DTZ Debenham Thorpe (1999) Research Department

Interest in car boot sales has also led to other research on the second hand market. The importance of this sector is often underplayed in retail analysis. More recent work on the role of markets, charity shops and newspaper advertisements remind us that for many residents of low-income households the second hand market is more important than new goods and services. For example, Williams and Windebank (2000) report on a survey of 200 deprived households on an estate in Southampton, United Kingdom. For comparison-shopping, they noted that only 46.5% of the sample would typically buy consumer goods from traditional shops (a figure that ranges from 62.5% for the most affluent within the group, to only 37.7% for the lowest income earners on the estate). For many households, obtaining goods from friends and relatives, markets and from answering advertisements in shops or newspapers was more important.

3.8   New Retail Formats and Distribution Channels

3.8.1   Introduction

In addition to new retail spaces, there are many examples of new retail formats or distribution channels. Indeed, it is increasingly argued that the task of network management is now shifting towards one of *channel* management as new forms of product and service delivery, such as telephone and Internet, emerge as both competing and complementary methods of connecting customers to suppliers. For example, financial services are now available through branches (and sometimes agencies), supermarkets, ATMs, kiosks (automated branches), the telephone and the Internet. In this section, we wish to explore some of the changes that are taking place within the retail market place and the threats they pose to the traditional branch distribution network. For each of these major trends we argue here, and in later chapters, that geographical analysis will remain crucially important in addressing these new distribution issues. The argument is that retailers need to develop flexible, local strategies for channel management based on a coherent unit of both customers and outlets (see Chapters 6 to 9).

3.8.2 Compact and convenient stores

Retailers are very resilient to change and difficulties. One of the major changes we have witnessed in the grocery market in recent years is the appearance of smaller supermarkets back on the high streets. This is partly in response to the new planning legislation that prevents out of town developments (see Sections 2.3.1, 2.4 and 3.3). In the United Kingdom, both Tesco and Sainsbury's have developed new formats. New store development continues in smaller or medium sized towns for a new range of 'Compact' and 'Metro' stores. By the end of 1996 to 1997, Tesco had opened more than 50 Compact stores (16 000 to 26 000 sq.ft) in towns that had previously been considered too small for a superstore. In Wales, for example, targets include towns such as Aberystwyth, Bangor, Cardigan and Milford Haven. In England, recent stores have been opened in Ashford, Faversham, Penzance and March. The Metro stores are even smaller developments associated with the traditional high street of many larger town and city centres. These are typically below 16 000 sq.ft. Recent examples include Manchester, Belfast, Richmond and Dundee. These Compact and Metro stores are not only inexpensive to build but also bring similar operating margins to the superstores. Thus, the return on investment of these formats is much greater (see the more detailed discussion in Guy 1995, Wrigley 1998).

Tesco have also targetted the growing convenience retail sector at petrol stations. As we discuss elsewhere, the convenience sector is now one of the fastest growing in Europe. Tesco's 'Express' format is based around convenience stores at the petrol forecourt. Latest openings in the United Kingdom include Leicester and Birmingham. They announced plans in 1998 to set up a nationwide chain of more than 100 Express stores in petrol station locations. Also, at the end of 1998, they announced a joint scheme with Esso to open Tesco Express convenience stores at Esso petrol station locations.

C-Store formats (under the brand names Tesco Metro, Tesco Express and Sainsburys Local), cater to a diverse customer base in local markets including pensioners, students, housewives and young professional workers who have a requirement to top up and, in some cases, even replace their weekly supermarket shop through a new channel. The first Sainsburys Local opened in Fulham in 1998. The store was small, only 3000 sq.ft, and thus offered only 2000 basic lines from the company's usual 23 000 products. The main emphases of the store are fresh produce and longer opening hours (6 a.m. to midnight).

It is useful to draw attention to the (re)emergence of the convenience store (C-store) format across Europe. The C-store concept has largely been drawn from the United States, where there exists a huge network of small stores selling products that are usually consumed within a relatively short period post purchase. In Europe, the C-store market is developing rapidly, usually through a franchise format (such as Spar, Alldays, etc.) whereby the C-store franchise operator has access to the buying power of the larger group. In the United Kingdom, Budgens announced in 2001 that it planned to open 20 to 25 new C-stores over the next few years, whilst the small C-store Jacksons announced plans to double the number of its stores in the United Kingdom by 2005 (mostly in the north of England).

Similarly, the new entrants to the UK grocery market in the early 1990s, such as Aldi and Netto, not only targeted the neglected deep discount sector but also targeted geographical locations in Britain's poorest urban suburbs (see Section 3.4). There are now other good examples of 'niche spatial marketing'. In Germany, the restrictions in building larger stores has led to the proliferation of a new wave of smaller, specialist shops. These are new self-service stores, specializing in a range of goods that previously could be obtained in discount markets and department stores (Vielberth 1995).

Table 3.5 summarizes the range of contemporary formats for Sainsbury's. The consumer is now served in many different ways by the leading grocery retailers; and, of course, it is not only the United Kingdom that now enjoys so many different formats. As Dawson (2000) notes, other European retailers are following this multi-format model of brand extension, such as GB, Casino, and Ahold.

3.8.3 Call centres

The main development in telephony as a distribution channel has been the creation of 'call centres', which link state-of-the art telecom technology with sophisticated customer relationship management systems. Caller information can quickly be accessed by the operator from the database and their account details and transaction behaviour received. The telephone offers two potential advantages: first, the ability to order *services* to be undertaken without the need to visit a branch. The financial service market has been quick to tap into this potential. First Direct, a subsidiary of Midland Bank, pioneered telephone banking in the late 1980s in the United Kingdom and others followed rapidly. Direct Line, which started life as a provider of motor insurance via the telephone developed a savings

Table 3.5 Sainsbury's store formats

Format	Sales Area Sq.ft. (approx)	Offer	Location
Local (pilot stores)	3000	Serving local communities: top up/grab and go	Towns, villages or railway stations
Central (pilot)	8–12 000	Catering for city centre shoppers' differing needs throughout the day	Major cities throughout the UK
Country Town	10–20 000	Small supermarket format–serving needs of weekly shoppers in small towns	Typically market towns throughout the UK
Supermarkets	15–30 000	Offering a wide range of products expected in a modern supermarket	Where possible, in line with planning policy
Superstore	30–65 000	Full 'superstore' offer demanded by today's customers	Where possible, in line with planning policy
Savacentre	65 000 and above	Complete superstore offer, plus hypermarket offers catering for the demands of today's shopper	Where possible, in line with planning policy

product in late 1995. In the first seven months of 1996, it generated £200 million of deposits, with a predicted growth of £600 million per annum (Reece 1996). Second, the telephone is increasingly used to order *goods* directly, especially price-conscious products such as car insurance. It seems clear that the inducement to shop by telephone comes from inexpensive products and better deals:

> Consumers who continue to use bank and building society branches to look after their financial affairs are losing out on cheaper loans and higher savings rates offered exclusively to the telephone based customers of the same financial organizations (Reece 1996).

There is an interesting geography to the location of call centres. Increasingly, these are located in areas of inexpensive labour rather than close to headquarters. India has emerged as a major call centre location, as has Scotland for many UK companies.

3.8.4   TV shopping channels and mail order

The growth of satellite and cable TV in Europe has allowed the introduction of US style shopping channels such as QVC and TV Shop. These channels specialize in lifestyle type products, such as kitchen utensils, fitness equipment, jewellery and so on. Owing to the low programme production and transmission costs they can devote several minutes air time to each individual product. Mail order is also a growing business expanding beyond the traditional 'multipurpose' catalogue into niche catalogues targeted at specific consumer groups. These have been dubbed 'specialogues' and include catalogues such as Kingshill, Mini Boden and Racing green. As Crewe (2000) argues, such niche catalogues have gone a long way in revamping the tired image of conventional mail order catalogues. Recent developments in digital printing will lead to unique catalogues being produced for separate households and individuals–true one-to-one marketing.

3.8.5   Automated branches

One of the major drivers of change in retail banking is the emergence of new channels for products and service delivery. Perhaps best known among these new channels is the automated teller machine (ATM) that has emerged as the most popular method for dispensing cash to consumers over the last 20 years. Figure 3.16 shows the growth of ATM usage worldwide, and Figure 3.17 shows the level of ATM provision growth and usage in Europe.

The automated retail branch is a growing retail phenomenon. Argos has created small order-only 'kiosk' stores in market towns. These kiosks hold no stock, but simply a catalogue to order from stores in nearby larger towns and cities, and have limited storage space where orders are delivered. Similarly, Nationwide Building Society is examining a new automated branch in areas too small for a traditional manned branch. Aylesbury has been chosen for the trial automated branch, which has voice and video links to head office, an ATM and a 'multimedia computer kiosk' (see Field 1997, p.150).

3.8.6   Multi-franchise outlets

Another major change in distribution is likely to emerge from retail outlets that currently enjoy a monopoly in terms of the goods they sell. This is the case with most banks, building

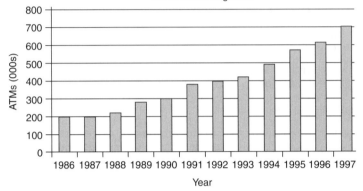

Figure 3.16   Worldwide ATM growth

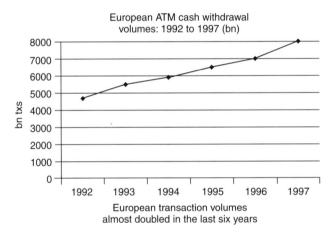

Figure 3.17   European ATM cash withdrawals

societies and motor vehicle dealers. The latter provides a good example of the change we might expect in the future. Presently, individual dealers, in the main, are only allowed by their manufacturers to retail a single brand. From a consumer perspective this means that brand comparison (for example between a Ford Focus and Vauxhall Astra) requires visiting two or more retail locations. In most of the retail sectors brand comparison takes place within the retail outlet. For the last 30 or more years, the automotive sector has been exempted from the Treaty of Rome that requires and allows retailers to offer multiple brand propositions to their customers. This so-called block exemption has been strongly protected by the manufacturers who argue that specialist dealers who are selling complex products with significant safety components require specialist and brand exclusive service repair facilities. The European Commission has, until recently, generally supported the manufacturers' position, but the review of block exemption in 2002 will no longer allow the manufacturers to insist that their dealers restrict their brand representation to a sole manufacturer's brand. It is likely therefore that we shall witness the development of a variety of multi-franchise retail formats, whereby consumers can compare different manufacturers' products in the same way they can currently compare washing machines, TVs and personal computers. Also likely is the emergence of department store type retail outlets that sell a variety of brands and the unbundling of a range of services currently offered by the traditional dealer (servicing, parts, body shop).

3.8.7   The home and office: E-commerce

However, the biggest challenge to the current retail market place is likely to come from e-commerce. Essentially, this involves some form of communication and/or transaction between customer and supplier through a computer terminal. One of the fastest growing mediums is the Internet. The German company Otto Versand, the world's biggest home shopping company, launched what it claimed to be the largest Internet shopping mall in Germany in 1997. It began with 13 companies, retailing 1.4 m articles for sale. Reynolds (1998) presents a useful discussion of the opportunities for e-commerce across Europe. He notes that despite overall impressive growth figures, the penetration of Internet use remains a 'patchwork quilt' affair (see Figure 3.18):

Strong growth comes from smaller European markets (most recently Norway and Belgium) as well as from the two largest markets, the United Kingdom and Germany. There is also some evidence of catching up from southern European countries (particularly Spain and Portugal). Yet, growth in France still remains disappointing, partly because of the dominance of Mintel (Reynolds 1998, p.5).

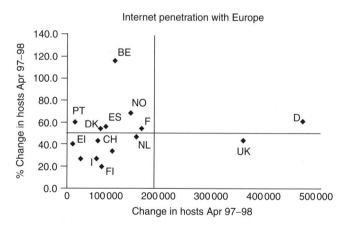

Figure 3.18   Internet penetration within Europe. Source: RIPE NCC

Another form of diversification that supermarkets are increasingly adopting is a provision of a direct home delivery service. Iceland was the first UK organization to offer a nationwide service. Tesco soon followed suit by announcing their home delivery service, through which customers can order primarily through the Internet, select foods from their entire supermarket range, and pay only a £5 premium for their basket of goods to be delivered to their home address at a time that is convenient to them. A slightly different approach is being adopted by Asda in the United Kingdom, where they are targeting home delivery services through a mixture of telephone and internet access in areas where they have little or no supermarket presence (for example, South London). The motor industry is also exploring e-commerce in a big way. Buying a car from a conventional dealership is often cited as more fearful than going to the dentist! Dealers, almost invariably, only sell one manufacturer's brand (though see discussion in the previous section) and shopping around therefore requires a consumer to visit several dealerships in different locations. Once inside a dealer's premises, an aggressive sales person, whose remuneration package is very much geared to the level of sales, often confronts the consumer. The tradition of haggling over the price of a new car and debating about the value of the trade-in often leaves a customer confused and irritated. It would therefore seem that an approach based on selling cars on the Internet would prove very attractive for a range of consumers. First, in the United States, but more recently in Europe, Autobytel has become the Amazon.com (the large book retailer) of the car-retailing world. Autobytel is essentially a new car brokerage service that allows dealers to bid for a customer. Autobytel signs up dealers who pay a monthly membership fee. A customer requests a quote on a certain brand and model line (e.g. Ford Focus 1.6, five-door hatchback) and this referral is then passed onto the appropriate dealers in the network. They then quote a price for the vehicle and Autobytel responds to the inquiry by passing on the lowest price quotation. The customer then makes contact with the dealer in the normal way. Autobytel can also arrange an appropriate finance package, insurance, and extended warranty for the customer.

In the United States, Autobytel has about 2700 car dealers associated with its network. It has recently launched similar services in the United Kingdom and Sweden. According to the market research company JD Power Associates, Autobytel currently leads, in terms of new car sales, with about twice the volume of its closest two competitors combined. It also ranks top in terms of dealer satisfaction and boasts the highest deal-closing ratio.

However, Autobytel is not offering a true e-commerce solution, but is effectively acting as a broker, introducing potential customers to a dealer who is prepared to make the best price offer to them. As such, it does not yet threaten the traditional dealer structure. Manufacturers enter into dealer agreements with their retailers, which prevent the manufacturer from supplying customers directly. The only company in Europe that sells directly through the ownership of its own retail network is Daewoo. Although the Internet is an ideal format for researching information about new vehicles and their price and specification, for most customers, the complexity of the transaction involving a trade of a used vehicle and the putting together of an appropriate financial package suggests that the Internet will remain a popular channel for information, but a niche channel for actual purchase. In the United States, it is estimated that only 2% of new car purchases are actually transacted on-line.

Given the hype over the future of the Internet and e-commerce, we explore this new distribution channel in more detail in Chapter 5.

66

3.9 The Old versus The New

It is our assertion that existing retail channels are likely to sit alongside the newly emerging ones rather than be taken over by them. However, retailers must learn to live with them both and adapt to take advantage of all types of format. In this section, we draw on our own pilot GMAP Channel Usage Index. This reveals not only very interesting up-to-date facts and detailed information about customer usage of distribution channels but also has far reaching implications for retail companies. The example we will use here draws on operations in the United Kingdom Financial Services Market. The study enumerates a number of key changes in customer behaviour in their use of different distribution channels and points out that companies operating in this area have not responded well to meeting those needs in the most effective way, both from the customer's and their own point of view.

The main highlights from a strategic viewpoint are:

1. Customers are becoming major multichannel users and unless steps are taken to prevent this, all the signs point to increasing channel usage by nearly all customer types. Without more radical action from companies in the financial services area customers will continue to use a wide variety of channels offered to them, and continue to drive up the total distribution bill for each company.

2. Far from disappearing, the branch role is still extremely strong. Many customers continue to view the branch as playing a critical role for them in delivery of financial services. The majority of customers are still fairly regular users of branches, with few indicating any significant change in this behaviour.

3. Branches are used by far too many low value/high transaction customers. On the whole, most banks have not been successful in wooing these customers to lower cost alternative channels.

4. In overall terms, banks demonstrate poor performance in attracting customers to use low-cost alternative channels to the branch. Fewer than 1 in 10 customers use alternative channels in the branch regularly and those that use them do so for collecting information rather than buying.

5. Use of the Internet is still very limited and is seen by most customers, at present, as an addition to branch usage rather than as a replacement for it.

6. Bank performance in converting branches to sales channels vary considerably. Some banks have been very successful in creating a high level of sales meetings, but others have branch networks that are heavily transaction orientated.

7. Many of the new channels, which banks have developed in the last few years, are simply not being used by customers–7 of the 15 channels reviewed in the survey have not been used by more than 1 in 10 respondents in the last three months.

8. Not only are a very limited number of customers accessing the Internet channel, but they are largely using it for collecting information and comparing prices rather than purchasing financial services products.

9. Although customer usage of new channels is limited, the take-up by different types of customer varies considerably. Substantial scope exists, however, for banks to target more specific customer groups by promoting the use of new channels; for example, the older age groups for internet delivery.

10. There are also considerable geographical differences in channel usage. This indicates that banks are not currently managing their distribution channel mix to meet the needs of their customers in the most cost-effective way. Perhaps, not surprisingly, Internet take-up is much higher in the southeast and London than elsewhere in the country (see Chapter 5).

The results of this survey indicate a strong need for more radical action to be taken by financial service companies in the distribution strategy area. There are five key areas that require more focus and attention:

1. Education and training of customers to use different channels.
2. Migration tactics and strategy for moving customers to lower cost channels.
3. Optimal distribution mix for each geographical market.
4. Focus on market segments and geographical markets requiring most attention.
5. Innovative and more radical solutions –Alliances/Partnerships, Branch Sharing, Franchising.

Let us look briefly in a little more detail at what might be done in each area:

1. Education and Training: Many customers are not utilizing new channels, simply because they are not knowledgeable about the way new channels operate and how they can access them. Banks need to invest more in demonstrating to key customer groups how to use a new channel and what the advantage of doing so are. Simply providing a new channel and expecting customers to use it will not work.

2. Migration Tactics: UK Banks and Financial Service Companies have been slow to use some of the pricing tactics successfully utilized by US and Australian banks to divert customers to lower cost channels. The use of pricing alone is probably not sufficient, but a more aggressive approach on utilizing a variety of different tactics is required to bring about a significant change in customer behaviour on channel usage.

3. Optimal Distribution Mix: Many banks have the same type of mix of channel for each geographical market they operate in. Some of the more advanced banks use geodemo-graphic models and other leading edge decision support tools, to help them build a different mix of channels for each geographical market they operate in. GMAP has already successfully developed the CMA (Customer Market Area) concept that helps banks address this particular issue. By matching the distribution mix to a particular market's individual customer needs, distribution costs can be better controlled and customer needs more readily satisfied. What is clear is that banks cannot continue to provide the same number and type of outlets in all the different markets they operate in.

4. Market Segmentation: There are considerable differences in customer channel behaviour between different customer types. Financial Services Companies will, in future, need to tailor their channel strategies to more specific groups. We are already seeing examples of physical branch outlets aimed at specific groups. In future, we are likely to see different pricing and promotional strategies for each market segment. This could also vary by geography as well as by product or social type.

5. Innovation: The need to drive down cost in the distribution area, but at the same time provide increasing customer convenience, is causing some banks to consider more dramatic action in the distribution area. We have already seen a number of examples of

Branch Sharing in different parts of the world and these are likely to increase further. Other areas being developed and explored include strategic alliances and partnerships with other types of retailers, franchising, automated branches, outsourcing, and more radical branch closure programmes. Although not all these strategies will suit every individual bank, we are likely to witness increasingly innovative and radical approaches to solving the key distribution dilemma, that is, how to provide customers with increasing convenience and access, but control and reduce costs at the same time.

3.10 Conclusion

In this chapter, we have looked at the future of traditional retail destinations and explored a number of new distribution channels likely to be more important in the future. It is clear that many traditional retail destinations will continue to be major retail growth locations, perhaps, with the exception of the superstore in many western European countries. Slowly, but surely, however, the new distribution channels will increase in importance until they eventually find some kind of natural equilibrium, when sales and market share will stabilize. Our argument is that retailers will need to be able to manage all these destinations and channels in the future. In geographical terms, this presents quite a challenge as retailers seek to find the right channel to serve the right location and the customers within them. Any competitive edge that a retailer can obtain in such an environment may prove to be precious. The pursuit of an optimal, intelligent location strategy has the potential to provide such an edge. Chapters 6 to 10 will aim to show the range of techniques currently on offer and how they can be effectively used in practice.

SPATIAL RETAIL GROWTH STRATEGIES

4

4.1 Introduction

The growth and expansion of retail organizations has always brought questions of geography to the fore. When a new retailer emerges on the scene, the location decisions can be quite straightforward – lets locate the first few branches in the largest cities. After that, it is likely that the location decision becomes more complicated and strategic. Although the 'new retail geography' plays down the importance of the traditional concerns with physical location and distribution, these problems have never gone away and remain crucial decisions for a retail organization. The methods for deciding on specific locations or location strategies are dealt with in Chapters 6 to 9. In this chapter, we wish to explore the options available for retail expansion and to provide examples of each strategy in practice. The interested reader should also look at a number of detailed case studies for individual firms (Laulajainen 1987, 1988, 1990, Graff and Ashton 1993, Sparks 1996). Finally we should also note that many of these growth methods are complementary, and many organizations will be committed to growth through more than one strategy. Cliquet (2000b), for example, examines the degree to which French firms have grown organically and through franchising. This is sensible, given the perils that growth by a single method may produce (see discussion below).

4.2 Organic Growth

Most retail organizations have used organic growth at some stage of their history. Organic growth refers to the process of opening new outlets (normally in premises owned by that organization, although renting is also a common option – see Guy 1994b). The speed of organic expansion is normally slow, but can be extremely rapid if finance is available. For example, Boots, the UK pharmacy, grew from its first store in Nottingham in 1877 to 550 stores by 1914. Guy (1994b) provides an excellent account of how such rapid growth can be financed. In many cases, extra shares rights can be organized, so that a lump sum is available for new properties. The growth of most large multiples can be categorized as falling into one of two types of organic growth. The first has been labelled hierarchical diffusion. This refers to the strategy of opening new stores in the largest towns and cities only, hence working systematically down the retail hierarchy. This is often also coupled with strategic regional development: that is, perhaps choosing a particular town from a number of similar options available (i.e. towns roughly in the same place in the hierarchy) as the basis of opening up new regional markets. Shaw (1992) provides the example of Liptons, an early player in the UK grocery market. Figure 4.1 shows the spread of Lipton stores in the United Kingdom by 1900.

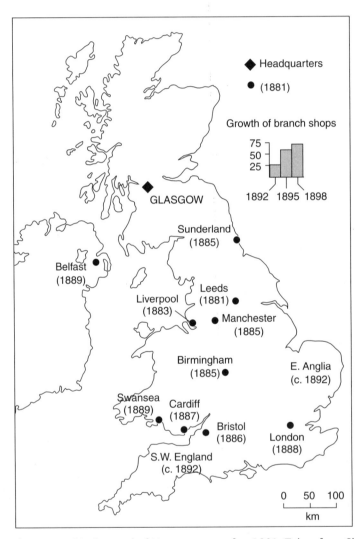

Figure 4.1   The geographical spread of Liptons stores after 1881. Taken from Shaw 1992

The second type of organic growth has been labelled contagious diffusion. This refers to the diffusion of outlets from a single location (usually the headquarters). This strategy first builds high regional or local market shares rather than national. Shaw (1992) also provides the example of Broughs in Northeast England (see Figure 4.2). Another good example is the early development of Wal-Mart in the United States of America. In the early days, Wal-Mart's aim was to become a 'category killer' in its hometown location. This involved

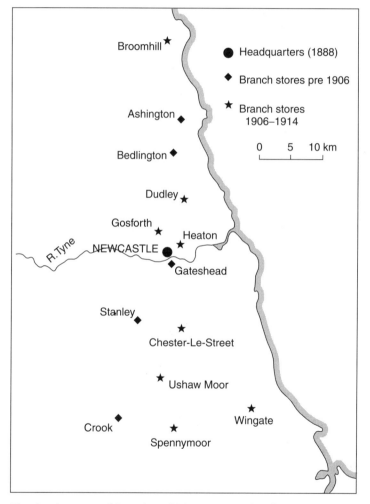

Figure 4.2 The development of Broughs, 1888 to 1914. Taken from Mathias, 1967 and Shaw, 1992

building as many stores as possible and offering such cheap prices that the majority of consumers converted to Wal-Mart and the competition was effectively killed (Graff 1998, Graff and Ashton 1993). Arnold *et al.* (1998) suggest that they could also be labelled as *market spoilers*, shifting consumer tastes towards their unique branding offer. Figure 4.3 shows how concentrated the pattern of Wal-Mart supercenters became in the United States during the mid 1990s.

Once an organization has reached a certain size, then their organic growth can be a mixture of the two. Alexander A. (1997) shows how Burtons first undertook a hierarchical diffusion strategy. Then, having established a number of key stores around the United Kingdom, they undertook a process of 'network infilling', building up more stores in between each of the core stores.

Organic growth is much more difficult in a foreign territory. Many retailers who use organic growth tend to adopt the process of contagious diffusion – opening new outlets across the border, where they can still be served from existing distribution points. The first non-UK European Marks and Spencer stores were in North-West Europe, close to the United Kingdom. This strategy allows the retailer to 'test the water', without investing huge sums in a new store network. If the overseas stores are not successful, then retreat is not too expensive, although it is likely to damage the reputation. However, expansion through organic growth is likely to be a slow process. Much faster geographical growth can be obtained through mergers or acquisitions.

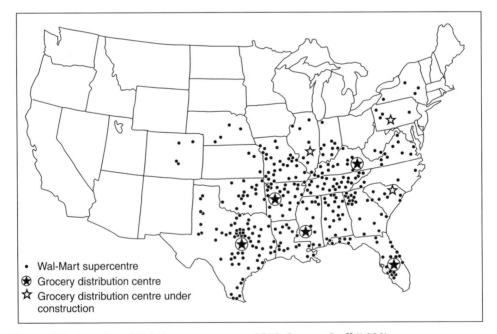

• Wal-Mart supercentre
✪ Grocery distribution centre
☆ Grocery distribution centre under construction

Figure 4.3 Location of Wal-Mart supercentres, 1998. Source: Graff (1998)

4.3 Mergers and Acquisitions

Mergers and acquisitions have always been an important retail growth strategy. Many organizations have used this strategy, not only to increase corporate power by eliminating competition but also to gain access to markets where they have been previously underrepresented. Such activity can produce rapid geographical growth, although there are risks and heavy costs associated with both strategies. Some recent European examples include Tesco's (who had no presence in the Scottish market) acquisition of William Low, Scotland's largest supermarket group; Casino's acquisition of Rallye; Albert Heijn (Ahold)'s acquisition of Primarkt; Rewe's acquisition of Billa; Netto's acquisition of the Ed stores in the United Kingdom; Swissair's acquisition of Allders International; and Jeronimo Martins's acquisition of Lilleywhites. Of course some mergers or alliances take place on a truly global scale. For example, Ford acquired Jaguar in 1990 and gained a controlling interest in Mazda in 1996. British Airways has a 25% share in Qantas and a partnership with American Airlines. In 1998, Glaxo Wellcome and SmithKline Beecham came close to creating the world's largest pharmaceutical company. Accountancy firms Price Waterhouse and Coopers and Lybrand merged on 1st July 1998, and BP and Arco announced merger plans in August 1998, creating an organization with a turnover of over £67 billion. In these and other cases, the quest is to guarantee economies of scale through reduced overheads and better product development. Burt and Limmack (2001) note that between 1982 and 1996 there have been over 1000 retail takeovers involving British companies. Wrigley (2000) provides an excellent table of major merger and acquisition activity in the global retail industry in the late 1990s – see Table 4.1.

The terms *merger, acquisition* and *takeover* have precise legal definitions that are classified on the basis of the number of shares one firm buys from another, the size of the firms involved and the accounting rules used. A merger is defined as the 'coming together' or combination of two or more firms to form a single new company. In a true, 'legal' merger, the two (or more) merging companies are subsumed – their shares being exchanged for shares in the new corporation. An acquisition is defined as the purchase of more than fifty percent of the share capital or assets by one firm (the acquirer) of another firm (the acquired), which leads to a change of control in the acquired firm (Hoschka, 1993). In a 'control' acquisition the acquired firm's shareholders cease to be the owners of that firm and the acquired firm – if retained – becomes the subsidiary of the acquirer. A takeover is similar to a control acquisition (i.e. one firm gains control of another firm through a majority stakeholding), but implies that the acquired company remains in existence. It also implies that the acquirer is much larger than the acquired (Sudarsanam, 1995). Such transactions have a cross-border dimension when the acquirer has its headquarters in a different country than that of the acquisition target.

The M&A planning process is typically described as having three main stages – the pre-acquisition planning stage, the negotiation and bidding stage and the implementation and integration stage (see Figure 4.4). The pre-acquisition stage is concerned with evaluating the best partner for merger within the context of a company's overall corporate goals and objectives. It can be subdivided into four main stages (as outlined in Figure 4.4). The starting point is an understanding of a firm's current competitive position and its strategic needs – strategic needs analysis. From this, a firm develops its corporate strategy and strategic objectives – that is, whether to move into a new geographical or product market, whether to strengthen its position in an existing market, etc. These strategic

Table 4.1 Food retail sector: merger and acquisition activity, Europe and North/South America: December 1997 to December 1999

Country	Deal	Country	Deal
USA	Fred Meyer/Ralphs merger	Germany	Wal-Mart (US) acquires Wertkauf
	Fred Meyer/Quality Food Centers merger		Wal-Mart (US) acquires Interspar
	Albertson's acquires Buttrey, Seessel's		Metro acquires Allkauf
	Albertson's/American Stores merger		Metro acquires Kriegbaum
	Kroger/Fred Meyer merger		Tengelmann acquires Tip
	Safeway acquires Carr Gottstein	Argentina	Ahold (Neth) acquires stake in Disco
	Safeway acquires Dominick's		Casino (Fr) acquires stake in Liberated
	Ahold (Neth) acquires Giant Food		Promodès (Fr) acquires stake in Norte
	Ahold (Neth) acquires Pathmark[1]		Promodès (Fr) acquires Tia
	Sainsbury (UK) acquires Star Markets		Casino (Fr) acquires San Cayetano
	Safeway acquires Randall's		Ahold (Neth)/Disco acquires Gonzalez,
	Delhaize (Bel) acquires Hannaford		Supamer
Canada	Loblaw acquires Provigo	Brazil	Ahold (Neth) acquires stake in Bompreço
	Sobeys (Empire) acquires Oshawa		Carrefour (Fr) acquires Planaltão, Mineirão,
France	Carrefour acquires Comptoirs Modernes		Roncetti, Raihna
	Casino and Cora create buying group		Sonae (Port) acquires Real, Candia, Big,
	Carrefour/Promodès merger		Mercadorama, Nacional, Coletão, Marmungar
Netherlands	De Boer Unigro/Vendex Food merger		Ahold (Neth)/Bompreço acquires PetiPreço
	(renamed Laurus)		Casino (Fr) acquires stake in Pão de Açucar (CBD)
UK	Somerfield/Kwik Save merger	Columbia	Casino (Fr) acquires stake in Exito
	Wal-Mart (US) acquires Asda	Chile	Ahold (Neth) acquires stake in Santa Isabel
Sweden/Norway	Dagab/D-Gruppen merger		Ahold (Neth)/Santa Isabel acquires Tops
	ICA/Hakon merger		
	Ahold (Neth) acquires stake in ICA		

Note: 1. Ahold withdrew from acquisition of Pathmark, December 1999, following opposition to acquisition from Federal Trade Commission.
Source: Wrigley (2000).

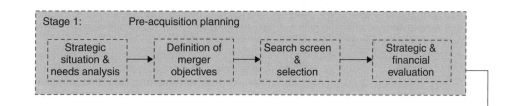

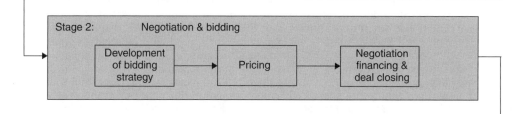

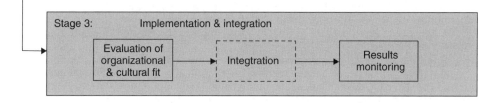

- - - - - - Included ——————— Not included

Figure 4.4 The M&A planning process (based on Sudarsanam, 1995)

objectives dictate the objectives of the merger (if M&A is the best strategy to achieve these strategic goals), which in turn dictate the search criteria for a potential partner. On the basis of these criteria, a partner is selected and evaluated against the merger and strategic objectives.

Having identified a suitable target, the negotiation and bidding stage is concerned with the technicalities of the transaction, that is, securing the acquisition of the target at the right price. It involves developing a bidding strategy (which is effectively a 'battle' plan, detailing the various lines of attack, outflanking manoeuvres and counter attacks), evaluating the value of the target and negotiating the price, obtaining the necessary finance and finally, closing the deal. This stage is predominantly external to the acquirer and target organizations; it is carried out by legal advisors and financiers (for more information see Sudarsanam, 1995).

The implementation and integration stage is concerned with the practicalities of bringing the two organizations together after the deal is complete. It involves a re-evaluation of organizational and cultural fit issues (given that at this stage more information is available than at the pre-acquisition stage) the functional integration of the various elements of the two companies involved (IT systems, personnel, distribution networks etc.) and the on-going monitoring of the performance of the new or combined entity (again, for more information see Sudarsanam, 1995).

The pre-acquisition planning stage is, perhaps, the most essential element of the acquisition process as it serves as the foundation for the process that follows (Hubbard, 1999). It is also one of the most overlooked areas of the process, and as a result, one of the main causes of poor merger performance (KPMG, 1997; Hubbard, 1999). (Recently Burt and Limmack (2001) conclude that while there is normally a boost in share prices prior to a retail merger or acquisition, in the post-bid period a significant negative pattern of abnormal returns is the norm). Although retailers may search for takeover targets that are a good business or geographical 'fit', in practice these choices may be constrained by non-spatial considerations, such as price, availability, ownership structures, product compatibility, management structures or simply the personalities of senior executives. Many of these factors become more important as retailers seek quick and defensive strategies of growth (Laulajainen, 1990). Nevertheless, we argue that the spatial dimensions, and more specifically, the potential performance of branch networks are critical factors. That is, in the retail sector, unlike in other sectors, the performance of a merger (whether a merger achieves its pre-defined objectives in terms of say market share or sales targets) is ultimately achieved through the aggregate performance of all the individual outlets acquired. Therefore, the performance (or potential performance) of the distribution network is an important (if not the most important) criterion in evaluating the best partners for merger.

As we noted earlier, prior to merger or acquisition, there is often little consideration of the ideal partner in terms of access to new geographical markets. There are notable exceptions. The Leeds brewer Tetleys used mergers and acquisitions strategically to undertake a process of contagious diffusion from Leeds, obtaining significant post war market shares in Sheffield and Bradford through acquisition, before merging with Walkers of Warrington to obtain high market share in Lancashire and Cheshire. Further mergers with Ansells saw them diffuse to the Midlands before acquiring national presence through an amalgamation with Allied Lyons. Another example is provided by the battle for Scotland in the grocery market. The two leading UK grocers, Sainsbury's and Tesco had little presence in Scotland by the end of the 1980s. As they perceived saturation to be more imminent in their homelands, Scotland became a fierce battleground. Tesco eventually won the battle in 1994 when they purchased the Scottish grocery retailing chain Wm Low & Co. plc. Figure 4.5 shows the spatial extent of the network thus gained by Tesco. The story of that battle, played out when Wm Low was itself trying to expand south, is well told by Sparks (1996a).

A third example is that of Carrefour in France. Burt (1995b) tells how in the 1970s and early 1980s they cleverly bought up parts of chains in different regions of France to increase their national coverage. Subsequently, in 1991, they bought out the troubled Montlaur company to give them greater presence in the southwest of France. This was followed by the acquisition of Euromarche in the same year, producing what Burt describes as 'a perfect geographical fit' (p.157).

Finally there has been the recent merger of BP and Arco. On agreeing to the merger, BP's Chairman emphasized how BP's high market share of petrol sales in Eastern United States of America would be complemented by Arco's high market share in Western United States of America:

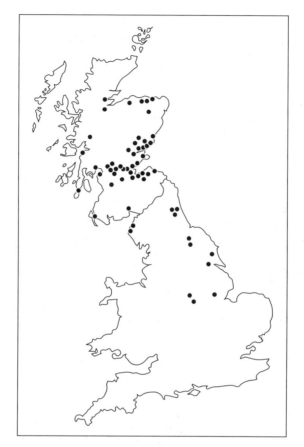

Figure 4.5 Wm Low store network, 1995. Source: Sparks (1996a)

For one thing it brings "coast to coast" coverage in the US refining and petrol retailing market, combining Arco's West Coast strengths with BP Amoco's east of the Rocky Mountains (quoted in Fagan 1999).

The importance of finding good geographical fits is also highlighted by problem acquisitions. Dixon's acquisition of Silo in the United States in 1987 for $384 million is a good illustration. By the early 1990s, Silo was operating at a loss, which in turn had a very negative effect on the Dixon group as a whole. According to Treadgold (1991), Dixons acquired a 'less than ideal store network'. This was because Silo was heavily located in the Midwest, an area of high and aggressive competition. Treadgold (1991) also reports that these areas were consequently overprovided with electrical goods space and returns were thus minimal. Dixons had seemingly made the cardinal sin of failing to carry out thorough research on the local markets within the United States (Docherty 1999) and they subsequently withdrew from the US market in 1993.

As noted above, geographical network planning is also useful in the integration stage. Once two branch networks are brought together by such circumstances, there is likely to be considerable overlap in their spatial representation, especially if they are national rather than regional players. The recent merger of the Leeds Permanent and the Halifax Building Societies is a case in point, with 78% of the original Leeds Permanent outlets residing in the same financial centres as a Halifax branch. In such cases, the gut reaction of the new combined organization is often to undertake closures *per se* – that is, to remove the overlap where a small town now has two branches of the same organization.

Such rationalization strategies are, however, often undertaken without detailed consideration of their impact on local market performance. A frequently pursued strategy is to simply close all outlets within a certain distance or drive time of an existing outlet. For example, Leeds/Halifax and Wells Fargo/First Interstate undertook such a simplistic strategy. More recently, both Barclays and Lloyds TSB have stated their intentions to take this course of action in their respective mergers with Woolwich and Abbey National. Barclays plan to close all Woolwich outlets within 100 m of their existing outlets and Lloyds TSB plan to close all Abbey National overlapping outlets within a quarter of a mile of their own. Such strategies, however, may not be the best way to proceed, producing little flexibility in very different geographical markets. As highlighted above, widespread branch closures after both the Leeds/Halifax and Wells Fargo/First Interstate mergers led to a reduction in revenues for the new combined organizations and an increase in market share for local competitors. Instead, we believe merger activity produces tremendous opportunities for the new organizations to review and revise their network distributions. Flexible, local solutions should be central to that review process.

It is in these aspects of the merger process – partner selection and network integration – that an understanding of the performance of distribution networks is a critical factor in ensuring success. Correspondingly, it is in these aspects of the process in which spatial analysis and modelling techniques offer huge potential to improve merger performance.

It should also be observed that in addition to the commercial and business drivers behind M&A activity, there is also a significant regulatory dimension at both the national and international scales. At the international level, a prime example is provided by current investigations by the European Commission, putting on hold planned developments in the business relationship between British Airways and KLM. Within the United Kingdom, the acquisition in 2000 of the Allied Domecq estate of public houses by Punch Taverns after a head-to-head fight with Whitbread is a prime example of a merger struggle that has been heavily influenced by regulatory issues. Furthermore, these issues have a strong geographical dimension, since the fundamental concern is whether the trade competitor (Whitbread) would be forced to sell many of the outlets acquired to avoid creating a monopoly position within local markets. Similar influences were almost certainly at play when the William Hill chain of turf accountants was acquired in 1998 by an investment bank (Nomura) rather than by its trade rival and initial predator (Ladbrokes). Another example is the planned sale of Lloyds TSB's mortgage brand Cheltenham & Gloucester. Lloyds TSB have been forced to submit such plans in order to allay the fears of the UK Competition Commission: that their proposed merger with Abbey National will not have an adverse impact on local competition within some areas of the UK mortgage market.

It is clear that geographers can play a major role by looking at the impacts of potential merger activity on market shares and the likelihood of concerns from organizations such as the Competition Commission (as it is called in the UK), and increasingly the European Commission (see Dawson 2000). Here we draw on the work of Poole *et al.* (2002b), as an example. The Competition Commission's report on UK supermarket retailing identified one hundred and seventy-five large stores (over 15 000 sq.ft.), where major retailers had effective monopoly or duopoly status in the local market. Despite the lack of immediate action to restrict future growth and concentration in the UK grocery market, the implication is that any attempt by the major parties at concentration via merger or acquisition will be firmly discouraged. Market analysts have commented that:

> any merger among the top five players would now lead to major store disposal programmes. For example, were Sainsbury to acquire Safeway, we estimate that around 40% of Safeway's selling area would be sold, at an unknown price, in order to obtain government approval
>
> (Deutsche Bank, 2000, 16 in Wrigley, 2001, 191)

In light of such an assessment of the UK market, this section examines several merger/acquisition scenarios, focusing on the concentration increases and possible sale of outlets that might result. Given a particular market concentration threshold, above which a retailer may be forced to sell off some outlets in order for a merger/acquisition to precede, which firms stand to benefit? Who is currently in a position to purchase additional floorspace in a local market without exceeding a particular 'threshold' market share?

Table 4.2 shows the increase in market coverage that potential mergers could produce. An alliance or partnership between Asda and Tesco, for example, would reduce the number of Postal Areas (PAs) where Asda previously lacked any penetration by twenty. The Asda/Tesco and Asda/Sainsbury's scenarios are referred to as an alliance or partnership, rather than a merger or acquisition, because Asda is unlikely to take over one of its larger rivals, despite the financial might of its parent, Wal-Mart. The retailer could try to enter into a strategic alliance with one of the top two, however, as a method of utilizing additional large stores needed to implement its strategies of diversified product assortments and greater economies of scale. Perhaps more significant than market coverage is the assessment in Table 4.2 of the number of markets (Postal Areas) where the floorspace share of the merger partners would exceed 25%. Although the combination of Asda, Tesco, Safeway or Sainsbury's with Somerfield always produces a significant increase in floorspace share (given the large number of Somerfield/Kwik Save stores), in reality, acquisition of Somerfield seems unlikely. The retailer is currently having problems integrating Kwik Save into its network, and there is also a disparity between Kwik Save's (soft discount) offer and the major player's offers.

The scenarios highlighted in Table 4.2 are, therefore, the ones producing the most significant increases in markets exceeding 25% share, excluding mergers with Somerfield. In the case of Tesco/Safeway, seventy-eight, some 63%, of Postal Areas would see at least one quarter of their grocery floorspace controlled by one dominant retailer. Tesco would control this proportion of the market in forty-six more Postal Areas than it had done previously. An alliance between Asda and Tesco would also give the partners over 25% of the floorspace in seventy-eight markets. An Asda/Sainsbury's alliance would

Table 4.2 Post-merger coverage a) PAs with >25%

Merger partners (major/minor partner)	PAs with no floorspace (post-merger)	No. change	PAs with >25% floorspace share (post-merger)	No. change
Tesco/Safeway	2	−5	78	+46
Tesco/Waitrose	7	0	43	+11
Tesco/Budgens	7	0	38	+6
Tesco/Somerfield	4	−3	90	+58
Tesco/Morrisons	5	−2	46	+14
Sainsbury/Safeway	1	−15	59	+46
Sainsbury/Waitrose	16	0	24	+11
Sainsbury/Budgens	16	0	22	+9
Sainsbury/Morrisons	14	−2	21	+8
Sainsbury/Somerfield	4	−12	71	+58
Asda/Tesco*	5	−20	78	+75
Asda/Sainsbury*	8	−17	50	+47
Asda/Safeway	3	−22	39	+36
Asda/Morrisons	23	−2	18	+15
Asda/Somerfield	5	−20	45	+42
Asda/Budgens	14	−11	5	+2
Asda/Waitrose	13	−12	6	+3
Safeway/Morrisons	9	0	11	0
Safeway/Somerfield	1	−8	56	+45
Safeway/Budgens	6	−3	12	+1
Safeway/Waitrose	6	−3	14	+3

*Alliance/partnership rather than merger/acquisition. Source: Poole *et al.* (2002b).

give Asda an additional forty-seven markets where it was associated with over 25% of the total floorspace. Sainsbury's/Safeway would have fifty-nine markets (almost 50% of Postal Areas) where it effectively had monopoly status, an increase of forty-six areas compared with Sainsbury's previous position. These figures are so high that it may be more appropriate to use the higher market share range used by the Competition Commission – over 40% – in order to undertake meaningful and manageable spatial analyses of merger scenarios.

Table 4.3, therefore, displays the same indicators as Table 4.2, for the 40% floorspace share threshold. Excluding scenarios involving Somerfield, the most significant increases in markets with over 40% floorspace share for retailers post-merger occur in the cases of Tesco/Safeway, Sainsbury's/Safeway, Asda/Tesco and Asda/Sainsbury's. Tesco/Safeway would be a particularly dominant combination, a merger or acquisition of Safeway by Tesco, resulting in the new retailer controlling over 40% of the floorspace in thirty-one UK Postal Areas (25% of total Postal Areas). A Sainsbury's/Safeway merger would give nine Postal Areas, where the resultant firm controlled over 40% of the floorspace. The equivalent figures for Asda/Tesco and Asda/Sainsbury's are fourteen and eight markets, with over 40% floorspace share, respectively. An Asda/Tesco alliance also produces a good spatial fit

Table 4.3 Post-merger coverage b) PAs with >40%

Merger partners (major/minor partner)	PAs with no floorspace (post-merger)	No. change	PAs with >40% floorspace share (post-merger)	No. change
Tesco/Safeway	2	−5	31	+26
Tesco/Waitrose	7	0	7	+2
Tesco/Budgens	7	0	9	+4
Tesco/Somerfield	4	−3	27	+22
Tesco/Morrisons	5	−2	6	+1
Sainsbury/Safeway	1	−15	9	+8
Sainsbury/Waitrose	16	0	5	+4
Sainsbury/Budgens	16	0	2	+1
Sainsbury/Morrisons	14	−2	3	+2
Sainsbury/Somerfield	4	−12	11	+10
Asda/Tesco*	5	−20	14	+14
Asda/Sainsbury*	8	−17	8	+8
Asda/Safeway	3	−22	6	+6
Asda/Morrisons	23	−2	6	+6
Asda/Somerfield	5	−20	11	+11
Asda/Budgens	14	−11	0	0
Asda/Waitrose	13	−12	0	0
Safeway/Morrisons	9	0	3	0
Safeway/Somerfield	1	−8	15	+12
Safeway/Budgens	6	−3	3	0
Safeway/Waitrose	6	−3	3	0

*Alliance/partnership rather than merger/acquisition. Source: Poole *et al.* (2002b).

in terms of merging Asda's northern heartland with that of Tesco's in the South. Only the peripheral areas of Scotland and mid-Wales would be areas of low combined market share. The Asda/Tesco scenario produces fourteen Postal Areas, where the retailer's combined share of floorspace is at least 40%. Two of these markets are located in Scotland, two in South Wales, three in South West England and the remainder in Greater London and the South East region. Currently, Asda does not hold 40% of the floorspace in any Postal Areas. If it were to form an alliance with Tesco, it would instantly gain over 40% of the floorspace in the SP (Salisbury), UB (Uxbridge), TW (Twickenham) and EC (East Central London) Postal Areas, where previously it lacked any penetration.

Figure 4.6 shows the geographical patterns of market share if the top two retailers in each postal area were to merge. It shows that between them, the top two retailers regularly account for 40 to 60% of the packaged grocery market, and often shows 60 to 80%. Areas of particular concern include the Home Counties (the area to the north and north-west of London), the south-east and south-west. The Scottish Highlands and Border regions are probably less of a concern given the smaller populations. Figure 4.7 shows the combined market share of Tesco and Sainsbury's in the United Kingdom. In many parts of the southeast their combined market share is regularly in the 60 to 100% category. A

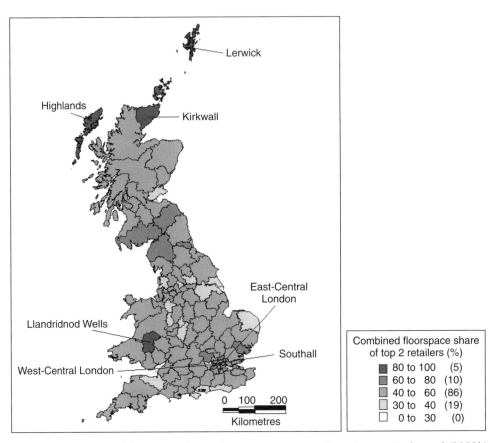

Figure 4.6 Combined floorspace share of top two grocery retailers. Source: Poole *et al.* (2002b)

merger/acquisition between these two organizations is highly unlikely, but, nevertheless, shows the market dominance of these big two in large parts of southern England. Poole *et al.* (2002b) offer a large number of what-if scenarios based on mergers involving many of the key players.

This analysis demonstrates that a merger or acquisition between the partners examined would significantly increase the geographical extent of the major players' market dominance. If such dominance is to be restricted in the future in the wake of the Competition Commission inquiry, it seems likely that potential partners could be forced to sell stores in particular markets to enable merger, acquisition, or partnership to proceed. The management of mergers and acquisitions is clearly a central management issue for the foreseeable future (Dawson 2000). Models to assess the implications of mergers and acquisitions will be discussed in Chapter 9.

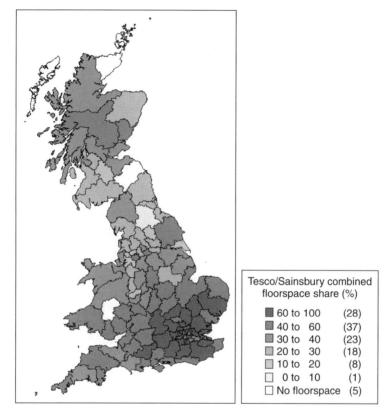

Figure 4.7 Tesco/ Sainsbury combined floorspace share. Source: Poole *et al.* (2002b)

4.4 Franchising

Given the concerns over competition and saturation outlined in Section 2.3, it seems that many retailers feel uneasy about opening and owning new stores of their own. This ties up capital and makes rapid expansion more difficult and expensive. An alternative is to get others to open and run shops on your behalf. This obviously reduces capital costs needed for expansion, although it means less of the profits come your way. This method of growth is known as franchising. A retailer (franchisor) offers their fascia to a third party (franchisee) in exchange for a lump sum fee and a return on profits. Global retailers operating franchise systems include McDonalds, Marks and Spencers, Body Shop, Benetton, Tie Rack, Burtons and Kentucky Fried Chicken. This method of growth has been especially popular in recent years for international growth. According to Burt (1995a), a quarter of all foreign retail investments by British retailers in the early 1980s involved franchising.

There are four main forms of franchising, but the most popular of these are direct franchising and the master franchise agreement (Quinn 1998). The direct method links the retailer or franchisor with the franchisee directly. The franchisor offers the franchisee the assurance of a tried and tested product, advice on operations and the products themselves (and possibly equipment if necessary). The franchisee offers the franchisor investment in the store itself, the ability to concentrate on product development, plus the knowledge that the franchisee will try to maximize income and profits, given that a share of those profits will accrue to the franchisee (unlike a tied or managed system in which the store staff normally would only receive a paid wage). With a master franchise agreement, the franchisor grants the master franchisee the right to subfranchise to others within an exclusive territory, thereby developing a three-way relationship (Quinn 1998). Thus, geographically, the definition of suitable territories becomes a major operation (see Chapter 6 for how this may be achieved). The advantage of this for the franchisor is that expanding the market further (often in unfamiliar territory for the franchisor) is in the hands of local experts. This is especially important for international growth. As Quinn (1998) argues, these middle organizations are usually in a stronger position to handle cultural and language barriers. A good example of growth by master franchising is provided by Body Shop (Docherty 1999). Through the master franchise system, Body Shop has been able to achieve a significant worldwide network (present in 45 countries in 1999), without sinking capital in a large store network. These arrangements are often strictly regulated, with the power lying in the hands of the franchisor. According to White (1995), for example, Next's requirements from its franchise partners are stringent, demanding that franchisees have expert knowledge of the fashion industry and that their stores be located on prime retail pitches (see also Docherty 1999). However, there is, perhaps, a risk that these middle organizations (that control all the franchisees for the franchisor) become too powerful and cause the original franchisor problems. Sparks (1995) tells the story of how 7-Eleven offered an agreement to franchise their name in Japan to the local Ito-Yokado chain. Such was their success franchising the 7-Eleven name in Japan that they grew large enough to eventually buy out 7-Eleven when it hit financial problems!

The details of the franchise approach can be found in Mendelsohn (1992a,b), Quinn (1998), Docherty (1999).

4.5 Joint Ventures and Strategic Alliances

In this section, we consider more formal partnerships between organizations that do not involve outright takeover or merger. These can be labelled joint ventures or strategic alliances. Although these terms are often used synonymously, there are some subtle differences. The term joint venture is now used to describe modest alliances between retailers – the promotion of one another's products and the building up of further scale economies through joint purchasing arrangements. The alliance between Kingfisher (UK) and Darty (France) in the mid 1980s is often quoted as a good example. Strategic alliances are said to refer to more formal arrangements. These involve fundamental decisions to share risk and are a key component of the corporate growth plans of both companies. A good illustration is provided by Harbison and Pekar (1998). They describe the strategic alliance between Wal-Mart and Cifra, the largest retailer in Mexico. When Wal-Mart began its international growth, there were many markets that it felt it lacked the expertise to trade from within. Mexico was a case in point. Wal-Mart felt it needed a strategic partner to understand trading in such a different retail market. Cifra was a good cultural fit for Wal-Mart, being a family-owned organization that shared Wal-Mart's belief in innovation, customer care and cheap prices. The benefit to Wal-Mart of the alliance was that Cifra understood the untapped potential of Mexico, and for Cifra there was a fast track along the steep learning curve of retail success:

'For Wal-Mart, Cifra would provide lessons in how to serve Latin markets... For Cifra the alliance was an opportunity to learn from the inside how the most efficient and profitable retailer in the World organized and conducted its business.' (Harbison and Pekar 1998, p68)

Such sentiments were backed up by Jaime Escandon, Cifra's treasurer (quoted in Harbison and Pekar 1998). He claimed Wal-Mart would gain from Cifra's merchandizing skills, purchasing skills, site selection strategies and inflation management. This seemed to be the case as Wal-Mart used the knowledge and experience gained in Mexico with Cifra to expand into many other Latin American countries. For a wider discussion of the pros and cons of strategic alliances, see Harbison and Pekar (1998), Doz and Hamel (1998), Nooteboom (1999).

4.6 Warehouse Location

In this final section, we reflect briefly on the importance of the physical distribution system for retail growth strategies. A retailer will not be able to maintain the business if he/she cannot get the goods to the shops at the right price. The siting of depots and distribution centres is an explicitly geographical activity and a classic geographical problem. However, before we elaborate on this, it is important to note that distribution today means much more than warehouses and lorries. It has been used as a term to describe not only the flow of products, but of risk, finance and information (Sparks 1998). It is also a term increasingly used synonymously with the term logistics. Here, however, we shall concentrate on the traditional concerns with transportation and the location of depots.

The dynamics of the distribution systems in most retail organizations is impressive. As we saw in Chapter 2, changes in the nature of distribution have been crucial for the leading firms earning greater profits. Since the 1960s, retailers have been increasingly more involved in the distribution system. In many cases, the entire activity of wholesaling has been internalized by retailers themselves (McKinnon 1986, 1989). Thus the middle organizations (linking suppliers to retailers) have been steadily eliminated from the supply chain. As retailers have taken more control of the distribution system, they have introduced fundamental reforms. The case of Tesco provides a good illustration. Smith (1998) plots the history of the Tesco distribution system. In 1989, 42 depots existed, serving most of England and Wales. Of these, only 26 were temperature controlled. At the same time, Tesco lorries were not compartmentalized and, thus, many single lorry trips were required to service all the stores for the entire product range. The whole system was too inefficient and there was too much variation in response time. What was required was a new smaller set of depots with much greater capacity for storing different types of goods. The lorries were also split into sections so that goods of varying temperatures could be handled in any one delivery. The net result was a new distribution system based on only nine depots, each serving around 60 stores.

The siting of depots is a geographical problem. Retailers find it extremely difficult and expensive to grow spatially without the necessary distribution back-up in place. This is especially true in an international context, but is true domestically as well. The expansion of Wm Low into England from Scotland was a good example of growth, constrained partly by problems related to distribution. Sparks (1996a) quotes the company as stating in 1993 that 'with the opening of (a branch in) Loughborough we have reached the southern limit of economic distribution from our existing depots and our future store development will not cross that line' (p1478). Thus the relationship between store expansion and warehouse locations is crucial. It is interesting to note that much research on optimal locations has been undertaken within the discipline of 'operations management'. However, we shall look explicitly at this problem in Chapter 6. For more discussion on distribution and logistics, see McKinnon (1986, 1989), Fernie and Sparks (1998), Fernie and Staines (2001).

4.7 Conclusion

In this chapter, we have provided a review of how retail organizations grow spatially. It has been shown that there are a variety of methods, each of which has its advantages and disadvantages. Organic growth may be the cheapest option in the short term, but this can be a slow growth method in this day of global corporate power. It may also be difficult in more competitive markets. Rapid geographical growth can be achieved through mergers and acquisitions, provided the finance exists (or can be generated). Franchising allows a retailer/manufacturer to achieve rapid growth while concentrating more on product development. However, the franchisor loses a great deal of control on how the product is retailed by the franchisee. Joint ventures and strategic alliances can reap considerable benefits to both parties, but they must be extremely carefully managed as they can allow one organization to come out of the deal more favourably than another.

It is likely that all retailers will be pluralist in the future, concentrating in the short term on more and better locations for their store network, and getting involved, from time to time, in the merger/acquisition market. Again, what is required is an intelligent approach to planning – not only appreciating the right strategy for different locations (domestic and international), but also planning effectively once a strategy has been decided. The latter might vary, from putting the right products into a new store opened organically to undertaking a major post-merger reorganization of two entirely different retail store networks. As ever, what is required are flexible, intelligent solutions.

GEOGRAPHY AND E-COMMERCE

5.1 Introduction

This chapter describes the development of e-commerce as a new channel that has been exploited by retailers, both existing players and new entrants. We shall also examine a number of issues of relevance, both in today's markets and in the future. These include:

Appropriateness and effectiveness of the internet in different retail sectors
Practical issues in e-commerce implementation
Pure Internet play–versus–multichannel approaches: who will win out?
Likely future developments

In many cases, we shall illustrate our discussion with examples drawn from a variety of sectors, such as supermarket and high street retailing, banking, automotive and other services.

There have, of course, been a plethora of books, papers, reports and guides on e-commerce in its broadest sense, and internet retailing specifically, and because of its dynamics it is almost, literally, impossible to keep up with everything! Even a widely admired text such as Michael de Kare-Silver's 'E-Shock 2000', first published in 1998 and revised in 2000, now seems out of date in certain respects (although still providing a treasure of historically important information and a wealth of case studies). So, as we write (September 2001), the only thing we can state with assurance is that a lot of what we project will probably turn out to be wrong! Other useful studies include Markham (1998), Jones and Biasiotto (1999), and Wrigley Lowe and Currah (2002).

5.2 What is E-Commerce/Internet Retailing?

There is a tendency to perceive Internet retailing as a single homogenous activity–the selling of goods and services via the worldwide web. However, there is a set of subtle differences between different types of retail activity via the net, and it is, therefore, useful to distinguish between the different types of services involved.

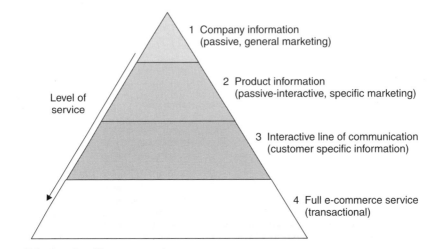

Figure 5.1 Levels of internet services

Figure 5.1 provides a useful way of differentiating between different levels of Internet use in retailing. Level 1 identifies those Internet sites that provide only general marketing information about a company and its products or services, with no connectivity, such as e-mail, providing a link between the customer and the company. In the mid-1990s' rush to establish Internet 'presence', many retailers' sites were of this type, but now they are relatively rare (at least in business–customer retailing). At level 2, we add the interaction, in the form of e-mail or phone-me capabilities, allowing the browser to make specific requests (e.g. send a brochure) or talk to someone about a specific aspect of a product or service proposition. At the time of writing, this would probably constitute the minimum Internet proposition a retailer would consider. Level 3 builds on level 2, but in this case captures information about the customer through some form of registration that is a prerequisite to gaining access to a retailer's site. Using customer specific information, a retailer can then target particular propositions to different groups of customers, via e-mail, post or phone. Level 4 services are the ones that have become most familiar in recent times, where the whole transaction is performed over the Internet, including payment.

Later in this chapter, we shall describe different retailers' approaches to e-commerce and the Internet. However, it is worth putting a few things in perspective with regard to Internet retailing. The significant hype that preceded the fall out in the dot.com sector during 2000/2001 disguised some fundamentals that can now be presented with hindsight. KPMG, for example, estimates that even by 2005, e-commerce will amount to less than 5% of all US Retail sales. In 2000, the total value of products and services bought by US consumers was $3 trillion, but the value of all goods and services bought on-line in 2000 was $12 billion–clearly a significant amount, but a small share. The day after Thanksgiving 2000, Wal-Mart's retail sales were $1.3 billion. In the whole of 2000, Amazon.com sales amounted to $1.5 billion.

5.3 Who Makes Money from the Internet?

With the collapse of dot.com companies over the last couple of years, it is easy to believe that no one makes any money from the Internet industry. A useful analogy is the Californian gold rush in the late 19th century. No one found much gold, but plenty of people made a fortune from selling pickaxes and buckets. Within the Internet economy, it is possible to identify four layers of distinct business that have different characteristics and profitabilities. Layer 1 consists of the infrastructure component of the Internet world. Within this layer reside the telcos, the Internet service providers (ISPs), the backbone carriers and those companies who provide end-user networking. In first quarter (Q1) of 1998, it is estimated that internet related revenues in this layer were $26.8 billion, in Q1 1999 this had grown by 50% to $40.1 billion.

Layer 2 consists of the applications infrastructure–software products and services that facilitate web transactions and transaction intermediaries. In this category we would include companies that develop Web sites, portals and e-commerce sites. Q1 revenues in 1999 grew to $22.5 billion, a 61% growth over Q1 1998.

Layer 3 Internet businesses consist of intermediaries who generate revenues through advertising, subscription fees and commissions. In this category fall companies such as Yahoo, E*Trade,, E-Bay and Last Minute.com who provide either web content or are market makers–putting prospective buyers in touch with prospective sellers. Q1 revenues in 1999 were $16.7 billion, a 52% growth over Q1 1998.

Finally, there is Layer 4, those companies who actually sell products and services over the Internet. Amazon is clearly the best-known example in this sector, but there are also prominent players, both established retail brands (e.g. Tesco) and new entrants (e.g. Virgin Cars). It is this fourth layer, which in revenue terms is the fastest growing, that has attracted most attention from the media, and is the one that demonstrated most volatility in terms of profitability and ability to 'burn' most investor cash without making any returns. It is interesting to note that those Internet retailers who have failed (and gone out of business) are, in the main, new brands.

5.4 Understanding Consumer Preferences

5.4.1 Introduction

One potential explanation for the demise of many retail dot.com businesses is their failure to understand the needs of customers in the market place. The technology surrounding e-commerce became so all-consuming that operators failed to undertake sufficient research on prospective consumers' reaction to it. Just because something is technically feasible, it does not necessarily make it desirable, never mind the basis for a successful business model.

To gain a better understanding of customer behaviour and attitudes in the retail financial services sector, in late 2000, GMAP undertook a detailed survey of a sample of UK current account holders (introduced in Chapter 3). In this section, we report on the findings of this survey, particularly as it relates to the use of the Internet as a banking channel, and also the implications the findings of the survey may have for retailers, both in the financial services sector and elsewhere.

5.4.2 The survey

GMAP surveyed 1000 current account holders in September 2000–the sample was weighted so that it properly reflected the characteristics of UK current account holders, in terms of demographic and geographical characteristics. The interviews were undertaken by telephone and lasted approximately 15 minutes. On average, customers had performed 17 transactions with a bank over the last three months. Because of the nature of the financial services sector there is a multiplicity of channels that customers can choose to use. Table 5.1 shows the 15 different channels we identified along with the percentage of customers surveyed who had used that channel, at least once in the last three months.

The first point to note is that the ATM and the branch are the dominant channels and that new virtual channels (but not the telephone) have yet to make much of an impact–we look at this in more detail later.

A second and important finding from the survey was that there is significant variation between banks in terms of Internet usage. While Barclays has the highest level of branch usage amongst the major clearing banks, it had the biggest share of Internet transactions, with HSBC a close second. Natwest and Lloyds TSB both appear to be lagging in the challenge to migrate customers to the Internet (see Figure 5.2).

In terms of the servicing of different products, different channels have varying degrees of popularity. Table 5.2 shows customer preferences for channel use by the major product categories. Not surprisingly, the branch and ATM dominate (although with different relative emphasis), but the Internet fares poorly, with only the servicing of savings accounts attracting more than 15% of customer preferences.

The results presented in Table 5.2 emphasize that customers prefer a multichannel world–one in which they can exercise choice and control over the purchasing and transacting behaviour.

Table 5.1 Channels used and customer usage

Branch ATM	Used cash machine at organizations own premises	66%
Remote ATM	Used cash machine located elsewhere	57%
Branch	Visit to branch/office to carry out transaction or make enquiry	49%
Telephone	Telephoned them	31%
Post received	Collected any information about them by a written enquiry/post	26%
Post office	Visited Post Office to conduct financial transaction/enquiry	20%
Meeting at branch	Had a pre-arranged meeting at office/branch	15%
Post sent	Sent anything to them by post	14%
Automated branch	Visited fully automated branch	9%
Internet home	Contacted/dealt with them through internet from home	9%
Kiosk	Visited branch/kiosk of financial organization in dept. store/shop	7%
Home meeting	Had a pre-arranged meeting at home/other place	7%
International work	Contacted/dealt with them through internet from work	5%
WAP	Contacted/dealt with them on a WAP mobile phone	4%
Interactive TV	Contacted/dealt with them through interactive TV	2%

Note: WAP-wireless application protocol.

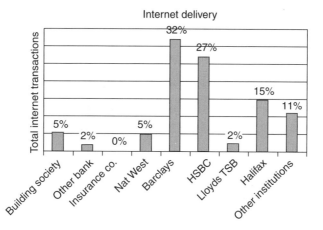

Figure 5.2 Share of all internet transactions by major provider

Table 5.2 Channel preferences by product

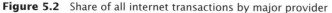

	Branch %	ATM %	Phone %	Post %	Internet %
Current accounts	31	62	4	1	3
Savings accounts	58	18	8	2	12
Credit cards	42	39	6	10	1
Mortgages	24	0	31	46	0
PEP/Mini-ISA	78	0	8	12	2
TESSA/Mini-ISA	81	0	16	3	0
Loans	32	0	21	47	1
Bldgs & Conts Ins.	48	0	23	30	0
Motor insurance	15	0	52	33	0
Life assurance	28	0	11	61	0
Pensions	36	0	34	24	7

A considerable amount of the Channel Usage Survey focussed on existing and future use of the Internet as a banking channel. Figure 5.3 examines current Internet usage. 50% of customers have access to the Internet, either at work or at home. Of these, about half (25% of the total) use the Internet for an hour or more each week, and 17% of all customers have used the Internet to purchase goods at sometime in the past. However, only 8% of customers have used the Internet for a financial service transaction of any kind, and only 1% have actually purchased a financial services product via the net (see Figure 5.3).

The reasons for this low usage are, no doubt, multifaceted and relate to some of the issues raised earlier; for example, the difficulty of acquiring relatively complex products over the Internet. To shed further light on this we asked customers about their level of comfort with different banking channels. The results are shown in Figure 5.4 and 5.5.

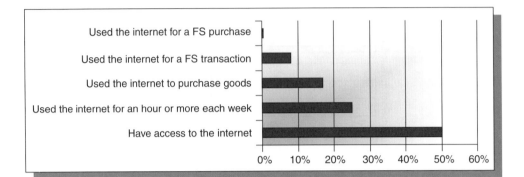

Figure 5.3 Current internet usage

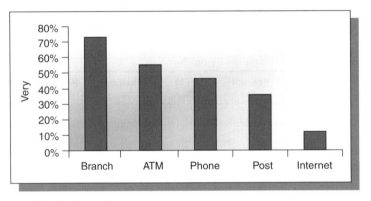

Figure 5.4 Comfort by channel

More than 70% of customers were 'very comfortable' with using the branch, but the Internet only scored 10% on the same indicator. Looking at this in more detail, Figure 5.5 compares 'total comfort' with 'some unease' responses broken down by different types of customers. Across all customers there is a 10/63% split between the two responses. However, the split changes when we look at net buyers, net users and financial service transactors. The conclusion appears to be that those using the Internet on a frequent basis are more likely to be comfortable with it as a banking channel–although even in the base case, total comfort levels fall below 50%. One may conclude that not only do internet sites need to be more user friendly and address issues of security, but that a significant education challenge has also to be met–to target those not using the internet, or using it infrequently to be persuaded of the benefits of the internet. One approach to this is to use bank branches as 'training centres', where customers are introduced to the benefits of Internet banking.

It is also interesting to examine the impact of the Internet on other transacting channels.

94

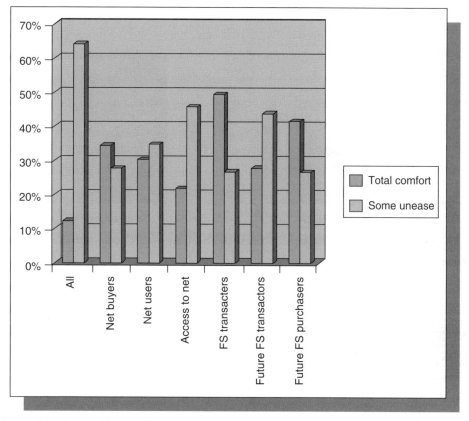

Figure 5.5 Comfort levels with internet usage

Figure 5.6 demonstrates that although customers using the Internet to transact make up only 7% of the customer base, they account for 14% of all customer transactions across all channels. When we examined the type of transactions processed through the net, we discovered some interesting results. Figure 5.7 looks solely at customers who are using the internet and breaks down the main transaction types into those performed on the internet and those performed using other channels. As we can observe, the proportion of Internet usage varies, with fund transfer, paying a bill, obtaining balances, and statements figuring high, but, notably, in all the transaction types there is a mix between the two channels. So, even those using the Internet to transact are still making considerable use of other channels.

Once again, this emphasizes our view of a multichannel world, with the need to provide Internet services as an integrated component of the product proposition, rather than a stand-alone product or service proposition.

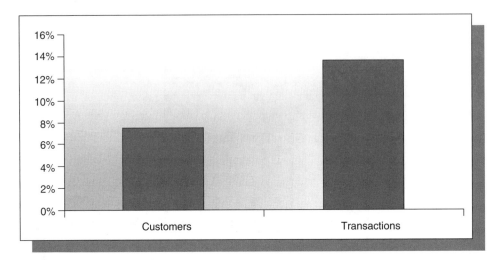

Figure 5.6   Internet customer profile

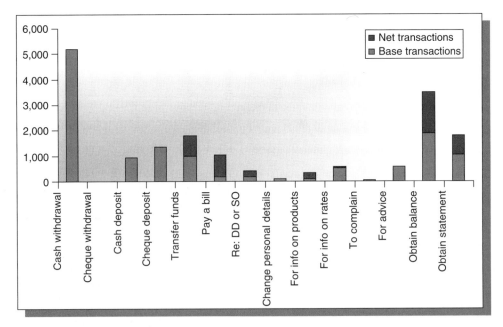

Figure 5.7   Mix of channel usage by internet transactors

5.5 Examples of Internet Retailing in Different Sectors

5.5.1 Introduction

In this section, we examine how retailers in different sectors have approached the provision of Internet channels in their businesses. We attempt to distinguish between the strategies that existing players have adopted and those of new entrants. We identify those factors that have enabled retailers to make a success of their e-commerce businesses and those that have failed.

5.5.2 Supermarket retailers

In most western economies, the grocery market is dominated by a small number of major players. In the United Kingdom, four groups account for over 60% of the market—Tesco, Sainsbury's, Asda and Safeway. None of these have been slow to attempt to exploit the Internet as a new channel for doing business. The customer proposition they have all adopted is pretty much the same, but the fulfilment strategy does vary. Contrasting Tesco with Asda will make this distinction. Tesco is probably the best-known Internet brand (it is also the market leader amongst the supermarket brands). Doing business with Tesco Direct is relatively straightforward. A customer accesses the Tesco Web site, registers and orders products from a long list of pull-down menus, containing more than 20 000 separate lines. The groceries and other products selected are stored as a 'shopping list', from which items can be added or deleted the next time the consumer wishes to make a purchase. The customers, having completed their selection, then selects a delivery time—essentially, a two-hour window–at most reasonable times and anything up to 28 days in advance. To complete the transaction, credit card details are provided by the customer, although the financial transaction is not enacted until the goods are delivered and the customer confirms the final basket price. There are a number of options for the customer, such as the ability to request alternative products if the selected item is out of stock.

Once the order is received by Tesco, it is forwarded to the supermarket designated to supply the delivery address. On the day of delivery, the order is picked from the shelves of the supermarket by a member of the staff. The items are scanned in the usual way and packed by the member of the staff. Frozen food, chilled and non-perishable items are packed in separate bags. A delivery van (operated by a subcontractor, but branded Tesco Direct) then drives to the customer's home at the required time. The driver helps to transfer the bags from the van to the customer's kitchen and gets the customer to sign the credit card authorization.

The customer pays £5.00 over and above the cost of the groceries for this service, irrespective of the value of goods purchased. This seems, on the face of it, remarkably good value, saving the customer's journey time, cost to and from the supermarket and time spent selecting the products from the shelves, queuing and packing the bags. The economics, from Tesco's perspective, seems bizarre. Assuming the average on-line transaction is £70.00 and working with their average net margin of 6%, this would generate £4.20 of profit on top of the £5.00 delivery charge. For £9.20 they have to pick, scan, pack and deliver the products as well as cover the overheads of the operation. This probably only makes sense if:

1. Staff employed to pick, scan and pack would not be otherwise engaged
2. Customers who shop on-line would not otherwise have bought from Tesco

Both the assumptions are probably weak, but it is generally recognized that Tesco are not doing this to make money, in the short term at least. They are doing it, like Amazon and other pioneers, in order to develop the expertise and experience so that as the market grows they become the dominant player, and make it difficult for new entrants to gain a foothold in the market.

Asda (owned by the American retailer Wal-Mart) has adopted a different strategy to home shopping. Its main aim is to penetrate those areas where Asda has little or no physical presence. Rather than fulfilling customer orders from local supermarkets, Asda have developed dedicated warehouses from which the orders are processed and delivered. The service is promoted and advertised in areas where competitors have the bulk of the market share. It is probably still too early to ask how successful this strategy is proving to be.

A third approach, that of Sainsbury's, is a hybrid of Tesco and Asda. Sainsbury's have constructed a state-of-the art automated pricing centre at Park Royal in London. Orders are processed here, but then despatched to supermarkets from where local delivery to the customer is organized and carried out.

These on-line shopping divisions of the major supermarket brands benefit from the advantage of the brand equity inherent in the parent company, the substantial financial resources the supermarket companies have, and their experience in supply chain logistics. However, the market share of the grocery market that on-line shopping attracts is still remarkably low. In 2001 Data Monitor estimate the total UK on-line grocery market to be worth £398 million, representing a 0.4% share. Tesco are estimated to have cornered more than half this market, with Sainsbury's and Asda both generating somewhere between 15 and 20% each. In the US, on-line grocery share of the market is estimated to be 0.33%, and in Italy it hardly appears on the radar at 0.02%. The ongoing price war in Germany has left supermarkets with little cash or appetite to invest in setting up on-line services. The highest penetration is achieved in the Netherlands, where it has reached 0.5%.

In the United States, we have seen the emergence of pure Internet play operators—new start-ups with no physical presence, but offering home delivery from warehouses in particular markets. Their fortunes have varied, but success stories are rare. Streamline.com, which installed fridges in customers' garages, went bust in November 2000. Peapod, with its stated aim to 'relieve Americans of the time and trouble of buying groceries', was bailed out in 2001 by the Dutch retailer Ahold in return for a majority stake in the business. However, the most spectacular failure has been Webvan who filed for bankruptcy in July 2001. Webvan was established in 1979 by Louise Borders, founder of Borders Books. Webvan sold food, non-prescription drug products and general merchandise over the Internet. It established a presence in seven US markets (Chicago, San Francisco, Los Angeles, Orange County, San Diego, Seattle and Portland) and generated very high levels of customer satisfaction amongst its 750 000 registered users. Many Americans found it almost too good to believe—15 minutes at the computer instead of an hour at the store, the savings in petrol and frustration were obvious—but the profit potential was not. Webvan provided a very high-level service—same day delivery within a 30-minute window, and groceries were delivered in special packaging. Delivery was free on orders of over $50. The company signed a deal with Bechtel, the construction firm, to design and construct its warehouses, with a roll out in 26 markets in a 2-year period.

Despite raising over $800 million in working capital, sales were disappointing. In the full year 2000, sales totalled $178 million, but losses ran at $413 million. Sales in the second quarter 2001 declined and eventually the money ran out. The failure of Webvan asks some serious questions about the potential profitability of home delivery of groceries. It would seem that customers are prepared to make use of services that provide a high degree of flexibility and do not charge a premium (or, at worst, charge a very small fee). However, it would appear that this business model is not a profitable one, at least for pure play companies. If you provide a less flexible service and charge a premium, then customers do not use it. This is why the existing supermarkets are in a strong position. They do not have to develop infrastructure in the same way new entrants would have to do, and they have larger purchasing power, and hence can squeeze margin out of their suppliers. The aspect of on-line grocery retailing that causes everyone problems is referred to as 'the last mile'–the delivery to the customer's home. In metropolitan areas congestion can play havoc with delivery times. In less well-populated areas, delivery just costs too much money. Ultimately, economics will probably win out, with customers having to pay for the cost proportional to the service benefits they receive.

5.5.3   Financial services

5.5.3.1   Introduction

We have already alluded to some of the issues concerning banking on the Internet in Section 5.4. In this section, we describe some of the approaches to Internet banking adopted by existing banks and new entrants. At the outset, it is important to recognize that the retailing of financial services is very different from other types of retailing.

First, there are many different products, ranging from the simple (e.g. a savings account) to the complex (e.g. a pension), but in most cases, there is no physical manifestation of the product (apart from cash). Consumers do not sit at home and admire their new investment account in the same way as they would a new car or hi-fi. This presents some obvious advantages to providing services over the Internet.

Second, the industry is highly regulated by government. In the UK, the Financial Services Authority imposes strict codes of conduct on banks and other financial service businesses to prevent scandals (e.g. pensions mis-selling) from happening again. This prevents certain products from being sold directly over the Internet.

Finally, for core banking products, such as current accounts and savings accounts, the product is purchased infrequently, but consumers rely upon a network of channels (both physical and virtual) to transact. Figure 5.8 shows the myriad mix of channels and products that a typical large bank will provide and operate.

In the rest of this section, we look at how different players have risen to the challenges of providing Internet services to their customers.

5.5.3.2   Internet strategies

It is useful to distinguish between two different sets of players in the financial e-world. The first group are those established banks who are either providing on-line banking for their existing customers or are developing new brands. The second, and smaller group, are those new entrants who are offering a mix of pure play internet services or are adopting a clicks and mortar strategy by linking up a provider with a physical network. As we shall

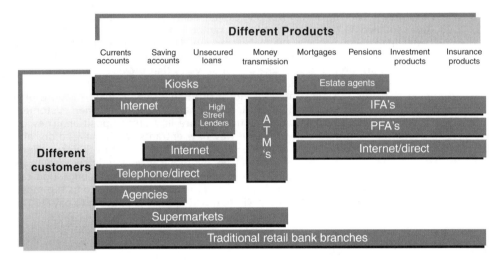

Figure 5.8 Channel mix for a typical financial services provider

discover, despite claims that barriers to entry for new players in this market were low, the majority of successful Internet banks are offshoots of existing providers who can leverage their substantial retail networks to support their customers.

It is also fair to point out that before the emergence of internet banking two very successful financial service businesses were built on the back of the telephone as the principal channel. First Direct, a subsidiary of HSBC, has built a base of over 1 million UK customers by offering a high level of service and competitive products. Although it now offers an Internet channel, the bulk of the businesses was developed on the back of a highly sophisticated call centre operation in Leeds. The bank made a strong case of its 24-hour 365-day a year service, and targeted customers whose lifestyles made it difficult to visit branches. Of course, it could call upon HSBC's own network of branches and ATMs for cash-based transactions.

The second example is Direct Line, a provider of car and household insurance in the UK market. Traditionally, insurance products of this type were sold through brokers or intermediaries, with the inevitable cost associated with this passed on to the consumer. Direct Line sold direct, over the telephone, and reduced the cost accordingly, providing an attractive proposition to the customer for a product that for many is seen as a distress purchase.

Established in the mid-1980s, within 10 years it achieved market leadership and spurned many imitators. Now a very high proportion of personal lines insurance is sold over the telephone, and increasingly the Internet. As price is one of the main factors in purchase and the product is relatively simple, car and home insurance are ideal products for the Internet.

In the mass markets of personal banking, virtually all UK clearing banks offer on-line banking facilities. Table 5.3 provides estimates of how many on-line banking customers the principal banks have. This should not be seen as a proxy for active users–they are

Table 5.3 On-line financial services customers

Country	Bank	Total customers (000s)	Telephone customers (000s)	% of total	Current PC/Internet customers (000s)	% of total	Target year-end PC/Internet customers (000s)	% of total
UK & Ireland	Abbey National	15 000	2000	13	0	0	1000	7
	Alliance & Leicester	5 500	1100	20	10	0.2	NA	NA
	Allied Irish Bank	800	70	9	43	5	100	13
	Bank of Ireland	600	150	25	30	5	250	42
	Bank of Scotland	5 000	1000	20	350	7	1000	20
	Barclays	13 000	1100	8	800	6	1000	8
	Halifax	20 000	1100	6	140	1	500	3
	HSBC (Excl. First Direct)	5 000	1000	20	20	0.4	NA	NA
	HSBC (First Direct)	965	965	100	200	21	NA	NA
	Lloyds TSB	16 000	1300	8	300	2	1000	6
	Northern Road	1 400	250	18	0	0	NA	NA
	Royal Bank of Scotland/Natwest	8 500	1100	13	460	5	NA	NA
	Woolwich	4 000	100	3	10	0.3	35	1
	Total UK & Ireland	95 765			2363	2		

Note: Total UK Banking population c40 000 (000).
Source: JP Morgan.

simply estimates of the number of customers who have registered for personal banking. As we reported in Section 5.4, only about 7% of current account holders transact on the Internet.

The main development in Internet banking in the UK has been the launch of new Internet banks, in the main, supported by a large parent. Table 5.4 lists some of these players along with the brand they have launched. The strategy that these Internet banks have developed has varied. Egg started by offering a single product–a savings account–at an extremely attractive rate. Customers recruited were then, at a later stage, cross-sold new products as they were released. This approach is a common Internet ploy–develop a loss leader to recruit customers and sell them additional products (hopefully at a profit) later

Table 5.4 New internet brands

Parent	Internet bank
Prudential	Egg
Abbey National	Cahoot
Co-op Bank	Smile
Halifax	IF
Virgin	Virgin One

on. Like most Internet banks, Egg has yet to make a profit, and senior executives have recently admitted that they feel they need physical branches to succeed.

IF and Virgin One operate a consolidated banking approach. Customers may have a range of savings and borrowing products (current account, savings account, mortgage, unsecured loan), but with Virgin One and IF they pay interest only on the net amount of borrowing they have. Although there are technical differences between the two banks' products they claim their customers can make significant monthly savings by consolidating their assets and debts into one account.

Examining the growth in customers that these new banks are projecting (and early evidence suggests that these are not being achieved), it would appear that while internet banking per se may be an attractive channel for certain types of customers (as part of a multichannel approach to banking), new banks are likely to develop as niche players rather than as major players. Banking products, especially current accounts, have a high built-in inertia owing to the complexity of switching from one supplier to another, and this makes persuading customers to transfer to a new supplier difficult. Regulatory changes will attempt to reduce this problem.

We now turn to a contrasting example, an attempt to build a Pan-European Internet bank. First-e bank was established by the French bank Banque d'Escompte and operated from Dublin, Ireland, employing some 280 staff. First-e was launched in September 1999, during the heady days of the Internet boom, when it looked like Internet-only banks would change the face of banking.

It provided Internet banking services to customers in France, Germany and the UK With low overheads. The strategy was to offer far better interest rates than the High Street banks. However, growth in customer numbers proved elusive. In the UK, it is thought they had about 50 000 customers, but less than 15 000 held a current account. As time went by, they recognized the weakness of an Internet-only approach, and there was talk of developing a modest physical presence, although this never transpired. On Friday, 7th September 2001, it was announced that it was closing its British and German operations, giving customers the option of transferring their accounts to the DAB bank or its UK subsidiary Self Trade. Did First-e bank fail because it was a victim of the dot.com fall out or was its business model wrong? Probably, the answer is a bit of both; as we have mentioned earlier, all Internet banks lose money in the early years, so the willingness of the parent bank to invest over this period is essential. However, the financial service market is extremely competitive and there are likely to be more Internet banking failures in the future. Even in the US, where we might expect internet penetration to be the highest, a survey by Gomey in 2000 concluded that US branchless banks have less than 4% of active on-line banking customers.

One area where the Internet is likely to play an important future role is in financial services data aggregation. People, typically, have a wide variety of financial records, often scattered across a range of financial services companies. For example, a consumer may have a current account and a savings account with Barclays, a mortgage with Halifax, a Tessa with Prudential, an American Express Card, an investment portfolio with Merrill Lynch and so on. Apart from getting monthly or annual statements, consumers find managing their financial affairs difficult. Financial services data aggregation brings information from many organizations and institutions together in one place, and the Internet is an obvious channel through which this can be enabled. With a web account aggregation service, customers provide the aggregation service with their different providers' account details, logins and passwords. The aggregators perform the tasks previously performed by the individual, 'behind the scenes'. The service logs into the various accounts on behalf of the customer and gathers the data. The customer can then view their accounts as unified reports and charts. The service provider will also provide prompts when certain bills need paying and when particular accounts need funds. The opportunities for the service provider are several: intimate knowledge of a customer's financial affairs will provide obvious marketing opportunities–both to sell financial services products and to advertise on the Web site other products and services tailored to that customer's profile.

5.5.4 Automotive retailing on the internet

5.5.4.1 Introduction

Car retailing in most western countries has, for a long time, been carried out through a selective distribution system. Auto-manufacturers have not historically owned their retail outlets. Instead, they have appointed dealers who are assigned exclusive geographical territories in return for meeting facility and other standards set out by the manufacturer. This results in a dealer having a local monopoly on the supply of a particular brand, and prevents other retailers from operating in the territory. One of the standards normally applied is that the manufacturer does not sell any other manufacturer's product from that site. This means that customers have to visit several sites to compare different products, something that rarely happens in other retail sectors. Furthermore, manufacturers also usually insist that each dealer site offers a full range of new car sales, used car sales and service and repair facilities. This leads to dealers requiring large sites, and economics forces them to operate from light industrial type locations as opposed to good retail locations.

In Europe, this distribution system contravenes the Treaty of Rome in that it is anti-competitive. However, the industry has been 'block-exempted' from the Treaty since 1984–arguing that special circumstances characterize the industry, such as the importance of safety in repair work, the need for specialized brand-specific product knowledge, and the requirement to recall vehicles to remedy defects. The next review of block exemption is in 2002 and the European Competition Commissioner, Mr Monti, has already indicated that the exemption, at best, will be diluted, at worst removed altogether. We could, therefore, witness a significant restructuring in the industry and the Internet may turn out to be a significant channel through which this happens.

It is worth pointing out that an automotive purchase is a complex transaction. A typical new car purchase will also involve the disposal or 'trade-in' of an existing vehicle, and settlement of the finance package associated with that vehicle, the creation of a new finance package, a new or revised insurance policy and the registration of the new vehicle with the licensing authorities. Dealers, experienced with reducing the difficulties in this process, are well positioned to effect the transaction. As we shall discuss later, the Internet is not well suited to handling this complex transaction and so there is a need to attempt to reduce the complexity. Also, there is the 'emotive' element of car purchase that involves seeing the vehicle in the flesh. After all, a car purchase is usually the second highest value transaction a household undertakes (after the house). It happens relatively infrequently (every 3–4 years) and involves a relatively complex set of decisions.

The current retail system, dominated by manufacturer appointed dealers, generates significant inefficiencies. It is estimated that somewhere in the region of 30% of the sticker price of a car is generated by post-factory gate costs–distribution, advertising, marketing, dealers and so on. Also, in recent times, there has been consumer uproar in the disparity in low prices between different European countries. Again, as we shall see, this has presented opportunities for Internet-based companies to exploit economic inefficiencies and price variation to the potential benefit of the consumer.

5.5.4.2 Retailer and manufacturer response

Since the emergence of the Internet as a channel for vehicle retailing about five years ago, manufacturers have been reticent to embrace the channel as a way of selling cars to consumers. The reason for this is simple. The agreements dictate that the manufacturers have assigned the rights to provide for all sales to be conducted through dealers–preventing manufacturers from selling directly to customers. Manufacturers were slow in providing Web sites that mentioned anything more than details about their products (level 2 services). However, other providers began to view the opportunity of exploiting this low cost distribution channel. In this section, we review the response from manufacturers, intermediaries, direct sellers and dealers to the Internet opportunity.

1. Manufacturers: As mentioned in the preceding text, manufacturers are constrained in selling directly to customers over the Internet owing to their legal agreements with dealers, but there have been many attempts to get round this. The Internet has proved to be an excellent showcase for manufacturers' products, with video clips, 3-D imagery and so on. A high level of content, referring to vehicle specification and pricing, is provided by all manufacturers, as well as details (and links to) to a customer's nearest dealer. Vauxhall, the UK subsidiary of General Motors, has developed an interesting strategy of selling direct to consumers. They have developed derivatives of various models that are not sold through the dealer network, but are available exclusively via the Vauxhall web site. Other manufacturers, notably Mercedes Benz, have got round the legal problems by acquiring their own dealer networks–key locations in London, Manchester, Birmingham and Glasgow. This, then, provides a route for them to sell directly, but to date they have restricted their activities to used car retailing. Over the next two years, we anticipate that many more manufacturers will commence direct sales as dealer agreements are reviewed in the light of changes to the block exemption.

2. Intermediaries: The Internet serves as an excellent mechanism for putting prospective buyers and sellers in touch with each other. In the automotive sector the best-known e-brokerage is Autobytel, a US company that now operates in Europe (the UK business is owned by Inchcape). Autobytel started life in the US as a way of generating customer enquiries for new vehicle purchase and passing these on to participating dealers who, then in turn, put a proposal back to Autobytel. The company then selected the 'best' proposal in terms of price and dealer location, and passed this back to the customer. If the customer wished to proceed with the purchase, then the transaction was concluded between the customer and selected dealer.

Autobytel generates its revenues not from the car sales, but from monthly fees that dealers pay to participate in the scheme. The more successful the site, the more dealers will want to be part of it. For dealers, it provides an additional source of business, effectively another form of advertising.

Since its inception, Autobytel has attempted to solve the 'complex transaction' problem identified earlier. In the United Kingdom, they offer a part-exchange service, vehicle finance (in association with Inchcape Financial Services) and insurance. Whether customers think that Autobytel have solved the problem or that they got as good a deal as at a conventional dealership remains in question. What is not in doubt, however, is that Autobytel is the most serious of the auto-intermediaries. In April 2001, it acquired its US rival Autoweb.com to create a company with revenues of more than $100 million per annum, 7000 participating dealers and over 2 million web site visits per month.

3. Direct sellers: There are a growing number of businesses that sell directly to customers over the Internet. In the UK, companies such as Virgin Cars, Broadspeed and TINS (a subsidiary of dealer group Pendragon) among many others sell direct. They too have attempted to make transactions easier by providing part-exchange and finance packages. Most of the independent retailers source their vehicles from dealers in continental Europe, where pre-tax prices are significantly cheaper than in the UK. However, dealers on the continent have to order Right Hand Drive vehicles and this highlights to manufacturers where the product is destined for. Manufacturers have to fulfil RHD orders (it would be illegal under European law to refuse), even if they charge a premium for this. Indeed, Volkswagen were fined 100 million euros in 1998 for refusing to cooperate with dealers.

The web sites operated by retailers are sophisticated in terms of their content and their ability to provide customers with quotations quickly. There may be some doubts about customers' willingness to buy from them. However, Virgin Cars, with the massive brand presence of its parent company, set a target of 25 000 sales in its first year of operation. It actually sold around 5000 vehicles.

4. Dealers: Dealers have not been slow to provide details on-line of their vehicle stock, both new and used. As with retailers, the Internet provides an ideal channel for providing content about cars in stock. However, most sites do not provide the ability for customers to purchase direct. One of the most highly developed approaches has been taken by Sytner, one of the leading retailers of prestige vehicles.

Sytner have developed an approach called C3. It is an integrated customer contact centre that allows a customer to access information by mail, phone or Internet about any product that Sytner sell across their dealer network. All aspects of a transaction can be performed by C3, except the conclusion of the deal, which Sytner insist must be undertaken in the dealership.

5.6   Geography, the Internet and E-Commerce

Before the dot.com collapse in 2000, many observers suggested that the ubiquity of the internet would render the importance of geography in retailing to a minor role. We argue, in this section, that in many ways the opposite has happened: the interest as a channel matures, and as e-commerce providers consolidate, the relationship between virtual and physical channels has strengthened, and understanding geographical variations in the use of the internet at a national, regional and local level has become more important.

Our argument is set out as follows:

1. There is substantial geographical and cultural variation in internet use and e-commerce activity. Despite the ubiquity of internet use, consumers in different countries have demonstrated different adoption rates–Table 5.5.

Table 5.5 Estimated European internet users

	1999	2000E	2001E	2002E	2003E
France	7.0	10.7	15.3	19.9	24.1
Germany	13.5	23.0	29.2	34.8	39.0
Italy	6.7	10.2	13.6	16.9	19.6
Netherlands	3.4	4.5	5.6	6.8	7.7
Spain	3.9	5.9	7.9	9.6	10.9
Sweden	4.2	5.1	5.6	6.2	6.5
Switzerland	3.9	4.8	5.1	5.3	5.5
United Kingdom	15.6	20.6	25.0	28.6	30.9
Total	**58.4**	**84.8**	**107.2**	**128.1**	**144.1**

Source: Jupiter.

*New access devices such as WAP will put the Internet in everyone's hands.

*Interactive digital TV will add to overall penetration.

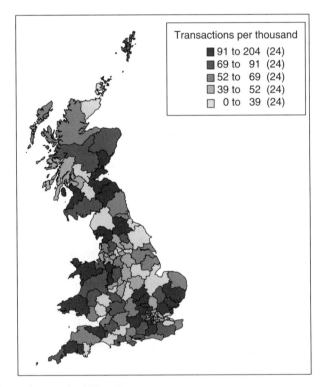

Figure 5.9 Channel usage by UK regions

Despite having similar populations and levels of income and population, it is estimated that France has half the number of internet users than the UK. At a regional level, Figure 5.9 Indicates how domestic e-commerce usage varies across Great Britain.

There are substantial variations between rural and metropolitan areas; perversely, in rural areas, where physical channels are at a premium, e-shopping occurs at a much lower rate than in areas with higher levels of provision. There will, of course, be confounding demographic and affluence effects here, but it is not what one might expect. At a more local level, Figure 5.10 presents our analysis of the estimated internet use for enumeration districts (ED)in Leeds. We concluded that there was something like a sixfold variation in internet use between the lowest and highest internet using enumeration districts. Again, much of this is explained by economic and geodemographic factors, but demonstrates the fact that there is considerable geographical variation in internet use and hence the ability of e-retailers to build market share.

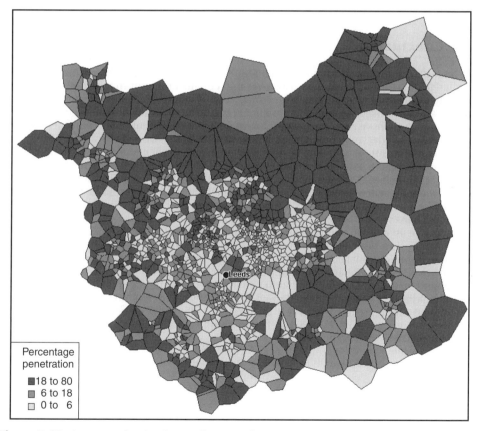

Figure 5.10   Internet adoption by small geographic area

2. Most retailers see the benefits of channel reinforcement strategies. As pure play, internet retailing in many sectors looks to have limited success and prospects, but as the e-commerce market matures the successful retailer of the future will be the one that adopts a multichannel approach, integrating and reinforcing the synergies among different channels. It is interesting to note that the most successful US bank on the internet, Wells Fargo, is still opening new branches in key markets and that 75% of its internet banking customers remain multichannel. In the book market, where internet offers a clear advantage in terms of price and ease of use, there have been more new book stores opened in the UK in the last 3 years than in the previous 10. These new stores have embraced the internet as a complementary channel, not a competing one.

So the euphoria of the dot.com revolution has been replaced by a new maturity. E-commerce is a channel that sits alongside others, but does not diminish the importance of geography.

5.7 Conclusion

It is clear that Internet sales are increasing and are thus becoming a more integral part of the channel management strategy. However it is equally apparent that e-commerce is still relatively small as a generator of sales in the retail sector. In addition, sales through the Internet are more likely to grow fastest in certain product areas – travel, leisure, books, flowers etc. There have now been a number of well publicised dot.com failures as retailers have failed to understand what consumers want or that they have anticipated much faster growth rates than consumers have allowed. Many solely e-commerce providers have now acknowledged that in the retail world physical branches are still crucial to financial success. In the financial service market for example, Egg announced in 2000 that they were considering a presence through a traditional branch network. At the same time, the Allied Irish Bank and the Sanwa Bank in Japan shelved their plans for e-commerce expansion quoting US banks that had reported disappointing growth.

The future is clearly multi-channel. Retailers will need to learn how to manage the Internet alongside both traditional channels (the branch) and any new future channels (telephone sales, automated branches etc). These channels must work together rather than as distinctive, often competing channels. To make that happen, retailers must understand that geography is crucial. As with all other channel formats, demand for e-commerce is not uniform spatially as is still heavily skewed towards certain geodemographic groups. Also, fundamental problems remain on the distribution side – how do retailers deliver a cost-effective dispatch system? The answer lies in optimising the channel delivery mechanism at the regional or local level.

TERRITORY PLANNING

6.1 Introduction

The concept of territorial organization goes back many centuries. For example, Jones (1976, 1985) has argued that the roots of Anglo-Saxon settlement patterns can be traced back to a network of 'multiple estates', in which many small settlements 'were subject to the jurisdiction of the territorial lord and, in return for their lands, paid rents in cash or kind and performed various services on his behalf' (Jones, 1976). This feudal system provides an early practical example of planning by geographical territories. At the start of the twenty-first century, planning the organization and distribution within geographical territories is still an issue of fundamental importance to the majority of contemporary organizations, for example:

a) *Political boundaries:* Numerous strategies have been devised over the years for the manipulation of election results to favour one party over another. Well-known examples include the practice of 'gerrymandering' constituency areas into infeasible shapes in order to manipulate shares of the vote, or by creating small constituencies in areas of political strength to be balanced by large constituencies in areas of political weakness. According to Johnston (2000b), this practice of 'malapportionment' has been outlawed in the United States of America (since) the 1960s, and British legislation requires the definition of constituencies whose electorates are 'as equal as practicable'. Such assignments may be supported by a form of territory management software known as a *districting algorithm* (Johnston, 2000a).

b) *Sales territory planning:* Any business with a direct sales component will need to organize its territories in an efficient way. For example, any pharmaceutical company will employ sales representatives in order to promote its products to family doctors or hospital consultants. Similarly, the manufacturers of branded goods will typically operate a sales force that is focused on the placement and prioritization of its products within retail outlets. These systems will work best when some kind of balance is maintained between the distances that representatives have to travel, and the number and type of meetings that they need to convene.

c) Franchise management: Owners of franchised brands face particular problems in territory management. Typically, franchisors are torn between the contrary objectives of assigning local control to one partner, and allowing maximum competition between franchisees within a local area. The classic example is automotive vehicles retailing. In the United Kingdom, each car dealership is allocated responsibility for a specific sales territory or DAR (Dealer Area of Responsibility), with many manufacturers employing a '20 mile rule', stating that no other franchise will be awarded within a fixed radius (e.g. 20 miles, or some other appropriate distance) of an existing sales outlet. In order to support this system, car manufacturers have been specifically excluded via a 'block exemption' from the regulations relating to freedom of trade in the Treaty of Rome. In the United States, USAI has built a multimillion dollar consulting business from a combination of sales territory planning and 'expert witness' services to support manufacturers whose desire to award a new franchise in 'Anytown' are challenged in the courts by the existing franchise-holder. Franchising is also a big issue for many High Street retailers (e.g. Benetton) and Quick Service Restaurants (see also Section 4.4).

d) Utilities: Providers of network equipment for mobile telecommunications face complex territory design problems. A typical objective is to provide services through a series of fixed points or 'cells'. In effect, each cell serves a territory and the objective could be to arrange the territories in such a way that the number of cells required is minimized. At the same time, sufficient capacity needs to be provided because once a particular cell is fully occupied, no new calls can be accepted. The problem has a number of additional complicating features. For example, a substantial proportion of the customer base is transient, as users may be connected from buses, cars or trains—these customers must be handed from cell to cell at appropriate times. Furthermore, the demand for services is difficult to estimate and continually growing. Hence, there is a problem of balancing what is required now against what may be required in future.

e) Warehousing and distribution: Organizations may need to supply customers inexpensively and rapidly according to variable and unpredictable requirements. For example, a car manufacturer might wish to provide after sales support to franchised dealers in such a way that customers can always access spare parts within, say, 24 or 48 hours. It is unlikely that such an objective can be met simply by maintaining an inventory of stock at a single warehouse. Therefore, the objective is effectively to design a series of territories once again, where each territory is served by a single warehouse. Note that in this situation, the objective is probably not to balance workloads between territories (as in the sales force planning example), but to minimize the maximum turnaround time to dealerships. This type of 'minimax' problem is familiar within the literature on location-allocation modelling (e.g. Ghosh and Harche, 1993) although good applied examples are hard to find. Similar problems are faced by retailers, who need to keep stocks within their stores continually replenished from the merchandize that is stockpiled for longer periods within their warehouses. This problem is no less acute to retailers at the forefront of e-commerce who will be increasingly required to provide products to their on-line customers within short timescales and at low cost. For example, the leading e-retailer Amazon has, so far, been able to build market share by providing very rapid delivery of books, compact discs and the like. This rapid delivery is sustained by an infeasibly expensive dependence on the postal service. At some stage, it will surely be necessary to serve its growing base through an inexpensive delivery network of warehouses or holding centres (see also Section 4.6).

In this chapter, we shall consider a number of problems that are relevant to retail businesses, and we provide examples in relation to each.

1. The simplest type of application is to map territories, and to provide measures of potential and performance within each territory. For many organizations, processes of this type are still effected using paper-based maps, even though Geographic Information Systems (GIS) provide a far more powerful and flexible means to the same end at modest cost. A review of the application of GIS to territory mapping problems is presented in Section 6.2.

2. In order to balance workloads between territories, some kind of automated design procedure will be required, such as the political districting algorithm discussed above. In effect, any set of territories represents a solution to this problem, and the quality of this solution can be measured by the degree of imbalance in the workloads. Therefore, what we have here is a simple optimization problem, in which the objective is to find a configuration of territories that minimizes the imbalance in workloads. Simple problems of this kind are discussed in Section 6.3.

3. Most practical problems of territory design will involve some form of trade-off between more than one objective. For example, in designing territories for sales representatives, the number of customers in a region may need to be considered alongside the travel time in reaching the customer, length of time to be allotted for the meetings and so on. Problems that involve weighting multiple criteria are considered in Section 6.4.

6.2 Territory Management

In this example, a supplier of stationery and office equipment ('Paper Planet') wishes to create a mobile delivery proposition that allows products to be distributed to small and medium-sized businesses. The proposition is to be developed using a franchising model, which means that each franchisee will have responsibility for customers within a specified geographic area. The business plan provides that franchise territories will be determined as geographic areas within 15 minutes drive time of a franchise location. In order to be viable, each franchise area needs to have a minimum of 2000 potential business customers.

There are two elements to the problem faced by Paper Planet. The first is to measure potential within a geographic area. The second is to manage the definition and allocation of franchise areas.

6.2.1 Measurement of potential

In order to support this application, a count of businesses across the United Kingdom is required. Data of this type may be obtained from a wide variety of sources, including directory sources (e.g. BT Business Database, Thomson Directories), database agencies (e.g. Dun & Bradstreet, 192.com, Blue Sheep) and government sources (e.g. Postcode Address File, Companies House). The most reliable data providers will typically collate and cross-check information from a variety of data sources, and will provide additional information about businesses from a classification of business activity (e.g. by standard industrial classification (SIC) code) to estimates of the number of employees, turnover or profitability. It is increasingly common for data suppliers to provide free access to limited information via their Web sites. For example, at the Companies House Web site (*www.companies-house.gov.uk*), it is possible to find the registered address and business activity of any named company within the United Kingdom without charge. Alternatively, BT's Yellow Pages Web site (*www.yell.co.uk*) allows users to access an inventory of all businesses of a given type within geographical areas of varying size, from a postal sector to a city or region. However, systematic access to data about companies of the type required by Paper Planet is only available by purchasing a database list, usually at a cost of several pence per thousand records. Companies House will provide detailed information on-line about directors, shareholders and trading activity of individual companies, but only at a cost of £5 for each individual company. This type of requirement would not be manageable for Paper Planet from either a financial or a logistical perspective. Fortunately, such detail is not necessary for current purposes.

For the present application, a measure of the total number of potential customers (i.e. businesses within an area) has been viewed to be appropriate. In a more sophisticated piece of analysis, it could be necessary to estimate the potential expenditure on office products within an area. In order to obtain such an estimate, it would be necessary to combine data about the number and type of businesses within each area, with intelligence of the likely spending of different types of businesses. Such additional intelligence would typically be gleaned from market research sources that could themselves be off-the-shelf or specifically commissioned in support of a Business Plan or strategic review. A further option would be to build measures of potential based on existing distributions of customers. For example, in the current application, if Paper Planet is an established supplier of stationery and office products, then it may have a database of existing customers that could be used to gauge the distribution of potential customers. This kind of approach is strongest in markets in which some form of syndicated customer data is available. A good example in the United Kingdom would be the car market, in which both the Society of Motor Manufacturers and Traders (SMMT) and Driver Vehicle Licensing Association (DVLA) can provide detailed small area data about the make and model of individual vehicles purchased within small geographic areas on a monthly basis. A broader review of the techniques that may be employed to measure potential within small geographic areas is provided in Chapter 8.

The final question to be addressed in relation to measurement of potential concerns the 'level of resolution' to be adopted, that is, what do we mean when we talk about 'small geographic areas'. In principle, it is quite feasible to address this problem with reference to individual locations. A good agency database will allow us to identify individual business locations at the level of an individual postcode, of which there are some one-and-a-half million. A similar level of resolution can be adopted in relation to the potential franchise locations. Thus, we could have a matrix of over 2 trillion possible customer-franchise relationships (1.5 million franchise locations multiplied by 1.5 million business locations). Although this is by no means impossible using current technology, it is wasteful and unnecessary. To understand why this is the case, we need to refer to the core objectives of the exercise, which are to define franchise areas of 2000 businesses within a 15-minute drive time. Neither of these components is a detailed empirical reality. 2000 businesses is clearly an estimate of the size of market needed to support each franchise. Similarly, 15-minutes drive time is both an arbitrary and fluid measure of proximity. 15 minutes in the evening rush hour is a very different proposition from 15 minutes in the early hours of the morning. Given that the parameters of our problem are sufficiently broad, there is nothing to gain by the introduction of very high levels of detail within the context of the application. In this application, we will, therefore, use postal sectors (of which there are approximately 9000) in order to define both the franchise locations and their territories. Thus, business potential is measured in terms of the number of businesses within each postal sector. This strategy has the desirable side effect that data at this level is much cheaper to obtain than at the level of individual businesses. Most importantly, however, this discussion can be seen as an application of the principle of 'Ockham's Razor': do not over-elaborate unnecessarily. A more detailed review of this principle is provided in the next chapter.

6.2.2 Area definition

Once the data on business potential has been gathered, the next step is to assemble the data within a Geographic Information System (GIS). While it is undoubtedly the case that some franchises are still being managed with the use of large paper maps, coloured pins and marker pens, it is also without doubt that such approaches are inappropriate in the context of twenty-first century information technology and analysis methods. GIS functionality allows the user to define territories, measure potential within those areas, to manage impacts such as the loss or relocation of a franchisee and to modify or create new territories.

For the purpose of the examples within this section and other chapters of the book, MapInfo has been used as the desktop GIS of choice for business applications. The functionality within ESRI's ArcView software is viewed as broadly comparable in scope and sophistication. Other products would be viewed by the authors as less effective, although potentially suitable for individual projects with specific objectives.

Following the preceding discussion, in order to support the application, two data layers are required within the GIS: one layer with business potentials for each postal sector, and a second layer identifying the drive time between pairs of sectors. Individual territories can then be created using the steps illustrated in Figure 6.1. First, a new franchise is created. Then individual postal sectors are selected and added to the new franchise area. The distance to each sector may be tested for violation of the fifteen-minute rule, and total business within the area can be displayed for accordance with the minimum business potential.

In Figure 6.2, we see how territories have been created in this way in order to cover the whole country, with large franchises in cities such as Manchester and Birmingham, and smaller operations elsewhere. It would be a matter of judgement for Paper Planet, whether to offer major city locations as a single franchise or to subdivide them to a greater number of franchisees.

There are many ways in which a set of franchise territories may be managed using a GIS application. For example, where neighbouring territories have differing potential business levels, then it may be thought desirable to reallocate postal sectors between territories, even if they end up being serviced from a greater geographic distance. In the case shown in Figure 6.3A, the East Blackpool franchise (the dark shaded area in the north west of the map) can be easily extended westwards (Figure 6.3B) to provide a better balance of potential with its neighbour in Preston.

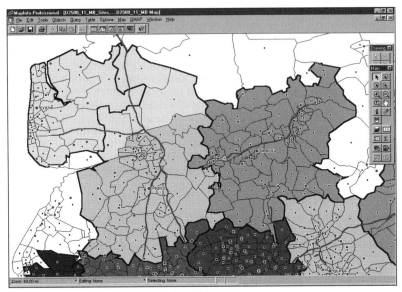

(A)

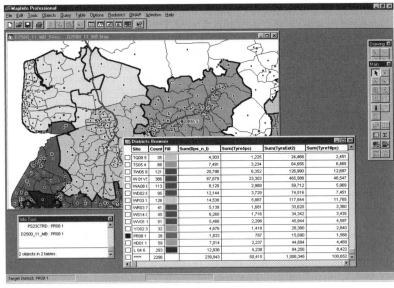

(B)

Figure 6.1 Steps in the territory creation algorithm

114

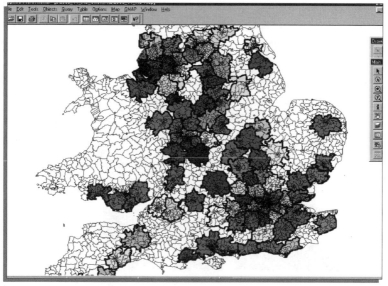

(A)

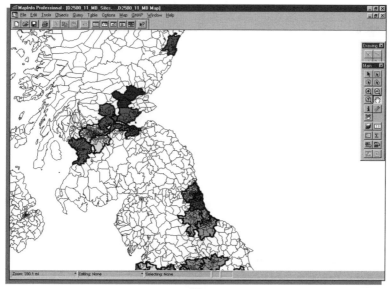

(B)

Figure 6.2 UK territories for the regionalization algorithm

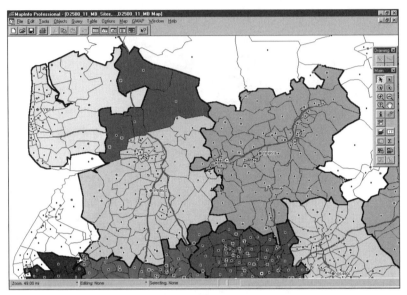

(A)

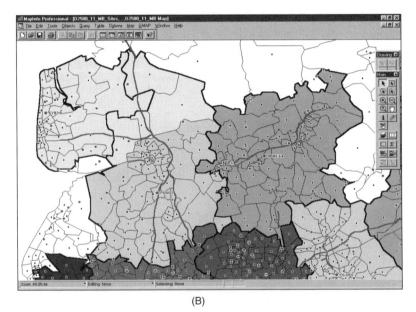

(B)

Figure 6.3 Territory allocation for the Fylde area of North-West England

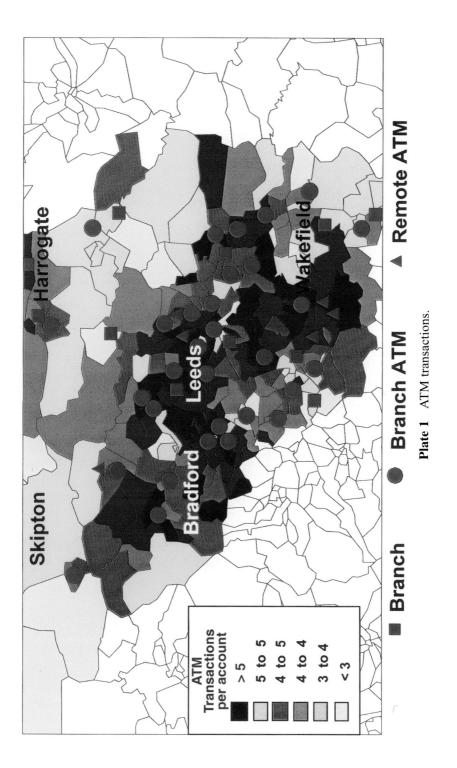

Plate 1 ATM transactions.

■ **Branch**   ● **Branch ATM**   ▲ **Remote ATM**

ATM
Transactions
per account

> 5
5 to 5
4 to 5
4 to 4
3 to 4
< 3

Skipton
Harrogate
Bradford
Leeds
Wakefield

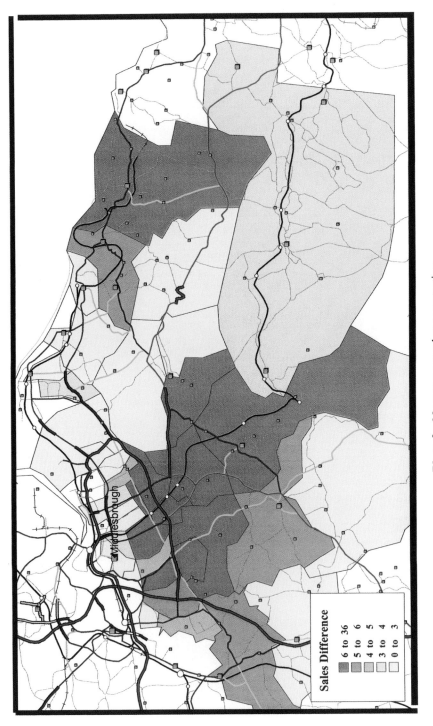

Plate 2 New store market penetration.

Sales Difference
6 to 36
5 to 6
4 to 5
3 to 4
0 to 3

Middlesbrough

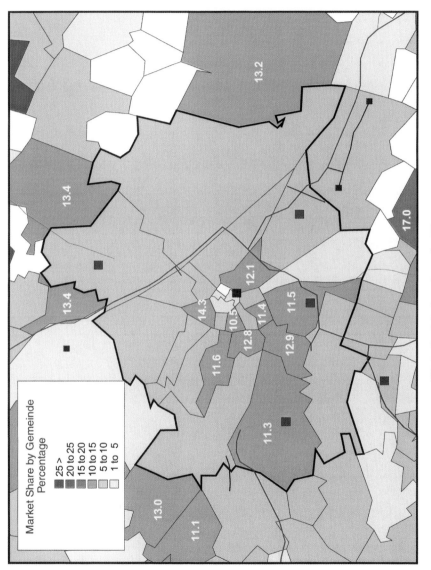

Market Share by Gemeinde
Percentage

25 >
20 to 25
15 to 20
10 to 15
5 to 10
1 to 5

13.2

13.4

17.0

13.4

12.1

14.3

10.5

11.4

11.5

12.8

11.6

12.9

11.3

13.0

11.1

Plate 3 Market share for Vienna.

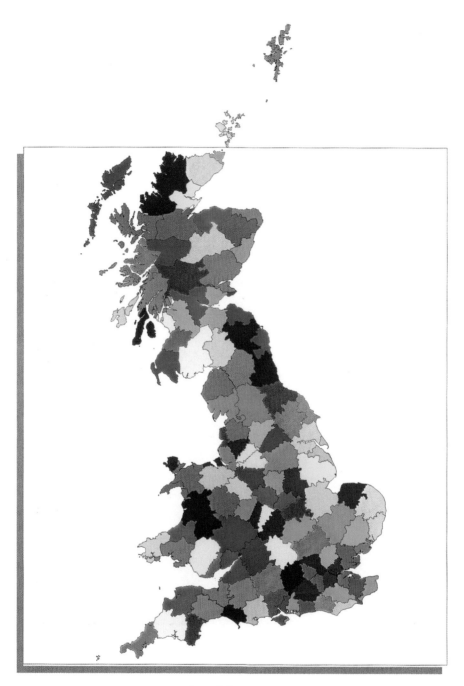

Plate 4 Driving test centres.

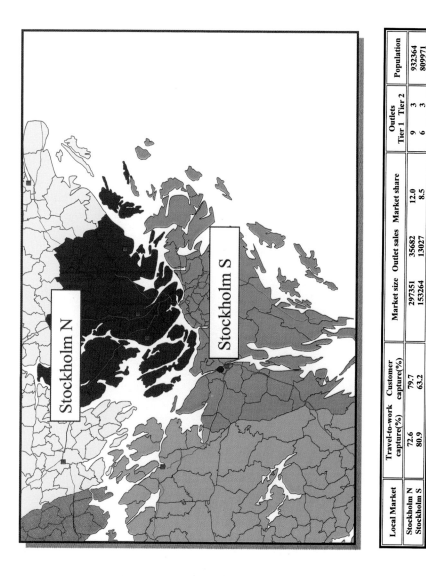

Local Market	Travel-to-work capture(%)	Customer capture(%)	Market size	Outlet sales	Market share	Outlets Tier 1	Tier 2	Population
Stockholm N	72.6	79.7	297351	35682	12.0	9	3	932364
Stockholm S	80.9	63.2	153264	13027	8.5	6	3	809971

Plate 5 Containment area solution for Stockholm.

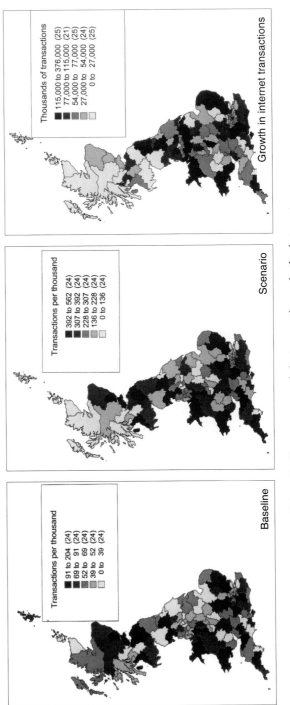

Plate 6 Forecast growth in transactions via the internet.

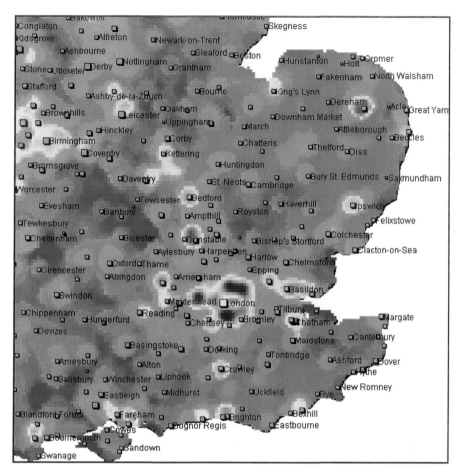

Plate 7 Supermarket potential map.

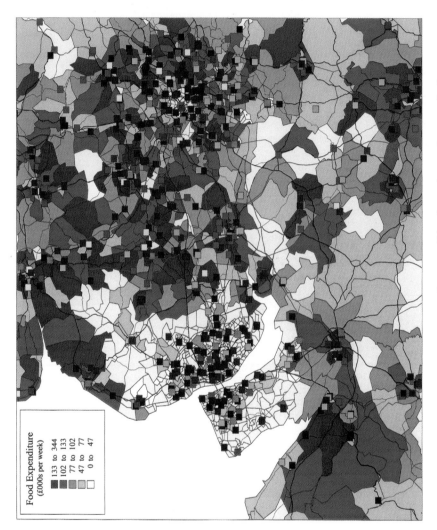

Food Expenditure
(£000s per week)

133 to 344
102 to 133
77 to 102
47 to 77
0 to 47

Plate 8 Market potential map for a retail business (second generation).

The whole of the Paper Planet application has been presented in the form of a 'business-to-business' (B2B) case study. It is, of course, possible to do exactly the same things for a business-to-customer application (B2C), such as a Quick Service Delivery franchise; or for a mixture of B2B and B2C, such as the type of photocopying franchise that can now be seen in the United States, United Kingdom and many European countries.

Once a franchise has been established, it may be possible to add further information to the GIS, which can help in the ongoing management of the enterprise. For example, a further overlay of customers could be created, which can be overlaid onto the franchise territories. This might then be used to generate benchmarking reports that facilitate performance comparisons between territories. It might also be desirable to compose more detailed reports about individual areas, for example, to show the mix of businesses, by size or activity type; or to look at competing suppliers or franchises in an area–either of which could have a significant impact on the ability of a franchisee to exploit his/her assigned territory. An example is shown in Figure 6.4, in which a territory has been built for the town of Toledo in central Spain, showing key socio-economic, demographic and business indicators for the area. In practice, such reporting capability is beyond the capability of desktop GIS, but can be linked in from another application, such as Microsoft Excel or CrystalReports. In the creation of a stand-alone GIS application for Territory Management and in the addition of further capabilities such as reporting, it is possible for organizations such as Paper Planet to develop their applications in-house, or to seek the assistance of

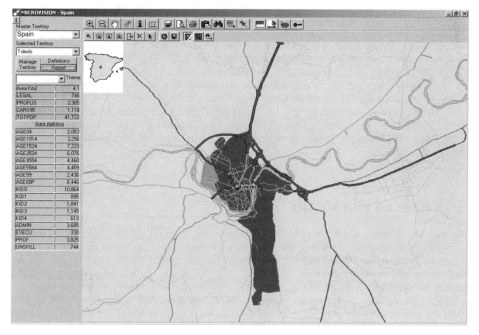

Figure 6.4   A sample territory report

external consultants. Both these approaches have their advantages and liabilities, but a fuller discussion is beyond the scope of the current chapter. More detail is provided by Birkin, Clarke, Clarke and Wilson (1996).

The usefulness of GIS in sales territory planning has been recognized by CACI in the construction of a customized variant of the well-known 'InSite' product, which is marketed as InSite*Fieldforce. The GIS capability of the system allows users to manipulate data and map existing territories (CACI, 2001). Users are able to make 'manual adjustments' to territories, 'in response to changes in the workload or field personnel'. It is likely that this functionality mirrors the examples discussed in the preceding text. The InSite*Fieldforce system allows users to create 'ideal territories', by assigning each area 'to the person who can reach it in the shortest possible time'. In promoting the benefits of this technology, CACI cite the fact that 'As a result the field force will spend less time driving and have more time available for calls. The solution will provide a significant saving on field transport costs and deliver productivity benefits at no extra cost in field resources.'

A major application area for InSite*Fieldforce is in the management of change. Considerations such as growth in the customer base, mergers and acquisitions, movement of personnel, and new product launches are all seen as destabilizing factors within the business environment. The software can be used to manage piecemeal revisions to existing structures. An important use for the GIS is to allow the production of high-quality maps so that, for example, a representative can see clearly how his/her territory has been redrawn, for example, in relation to existing customers, towns, transport routes and natural geographical features. Another benefit of the technology is the ability to produce analysis very quickly. For example, a territory planning exercise was undertaken for Coca-Cola and Cadbury Schweppes Beverages Limited on a time critical project. A structure of 600 new territories across nine regions was created as the basis for five business planning functions in only twelve weeks. Other customers of InSite*Fieldforce include Mitsubishi and Hallmark Cards.

In the Paper Planet application, we have assumed that individual territories can be created 'by hand' using GIS functionality, and of course this is perfectly feasible. However, we have also seen immediately, in our Blackpool-Preston example, that the problem of balancing potentials between territories will quickly become an issue. In order to achieve such balance effectively, something more than a GIS is required. It is to this problem that we now turn.

6.3 Balanced Workloads

In the previous section we have seen that it is possible to build up sales territories with the benefit of GIS technology, but that the solutions yielded by this process may be described, at best, as 'pragmatic'. In particular, we began to see that if some kind of equalization between the territories is desired, then a large amount of manual adjustment would be required to generate a good assignment of territories. A relatively complex version of this problem would arise from the Paper Planet application if it is required that as many franchises as possible be assigned, in addition to the existing constraints (that is, at least 2000 customers within 15 minutes). In the cities, this would then mean that we have to carve out franchises as close to 2000 customers as possible in order to maximize the potential for more franchises. Before reverting to this question, we first consider some of the general strategies that may be adopted to solve problems of this type, and then apply an appropriate method to a simpler problem, which is, nonetheless, of considerable practical significance.

An early solution strategy for problems of this type was provided by Openshaw (1977), whose Automatic Zoning Procedure (AZP) uses an iterative relocation method to optimize territories.

The problem of territory planning is highly complex and non-linear. One of the implications is illustrated, in a somewhat simplistic way, in Figure 6.5. Here we see a mathematical function that needs to be optimized with respect to some parameter. For example, this mathematical function could represent our desire to balance workloads, where the 'parameter' controls the assignment of territories. (In practice, therefore, this is not a single parameter but a complex set of spatial relationships—hence, this illustration is a simplification in two dimensions of a complex multidimensional reality.) Because the function is non-linear, it has more than one maximum value (1,2,3,4 in the illustration), of which only one represents a global optimum (3). The use of a hill-climbing solution strategy implies that a starting point is adopted somewhere on the solution curve, and is moved upwards from there. Hence, the maximum value that is reached depends on the starting point that is adopted. From Figure 6.5, starting values in the range A–B will tend to converge on the global optimum—others will not.

An approach to sales territory planning, using the concept of simulated annealing, has been proposed by Hurley and Moutinho (1996). The method is believed to be more effective than straightforward hill-climbing procedures, because although upward steps are always enforced, the possibility of downward steps is also permitted. In this way, it is hoped that local peaks to the optimization can be avoided. The scheme by which uphill steps are always taken, with occasional downhill steps is known as the Metropolis Algorithm (Hurley and Moutinho, 1996, p. 150). The method of simulated annealing 'has a direct

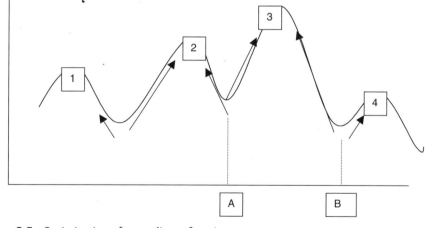

Figure 6.5 Optimization of a non-linear function

analogy with thermodynamics, specifically with the way that liquids freeze and crystallize, or metals cool and anneal.' The probability of taking downhill steps can be likened to the process of annealing at different temperatures. What usually happens within simulated annealing is that the 'temperature' is systematically reduced to force the procedure to a single maximum value, hopefully the global optimum. The proposal to use simulated annealing in territory planning is elegant and, most certainly, feasible. Unfortunately, no practical examples of its implementation are provided by Hurley and Moutinho.

The solution strategy that has been adopted in the examples contained within this chapter are more similar to Kohonen's concept of a 'self-organizing map' (Kohonen, 1984) than to either the AZP heuristic or simulated annealing. A description of the self-organizing map is provided by Openshaw and Openshaw (1997), together with an example of its application to the problem of clustering data for geodemographic analysis. The application involves eight steps:

Step 1. Initialization. The geometry of the problem is specified at this stage, that is, what variables are being used, how many clusters are to be identified, how should the data be standardized. These clusters can also be referred to as *neurons*–the self-organizing map is an example of an unsupervised 'neural network', to the extent that there are similarities between the way in which the neurons are guided towards good solutions to mathematical problems, and in the way that neurons in the human brain learn to solve real-world problems.
Step 2. Construct a random set of weightings for each variable on each neuron.
Step 3. Apply selective adjustments to any cases that may be affected by random noise, for example, outliers (cases that tend to resist classification) or cases with extreme values (e.g. small numbers).
Step 4. Assign data cases to neurons on a 'nearest neighbour' basis.
Step 5. Update the weights in the 'neighbourhood' of the winning neuron. The object of this step is to mediate in a process by which all the neurons will eventually become 'winners', so that their importance is somehow equalized.
Step 6. Successively reduce the learning parameters in very small increments. The objective of this step is to move the algorithm towards a stable solution. In the early stages, relatively large changes in the weightings are allowed; later on, these are reduced until the possibility of change is completely eliminated.
Step 7. Repeat steps 3 to 6 many times until convergence is achieved.
Step 8. Label the neurons and examine the self-organized map.

In passing, it is interesting to note that self-organizing maps have, in the past, been applied to problems in which the 'map' component is interpreted in a very broad sense. Here we are concerned with the organization of territories on maps within the narrow geographical sense. To the best of our knowledge, there are no published examples of the application of self-organizing maps to real geographical maps.

Now consider a relatively simple applied problem. The pharmaceuticals company, Spiridon, is about to launch a new palliative cancer drug. Its sales process will be supported by a field force of eight sales representatives. Each representative will have responsibility for consultants within Key Cancer Centres (KCC) located around the country. There are approximately 80 KCCs and 337 consultants. The objective is to find an allocation of sales representatives to KCC in such a way that each representative has responsibility for an equal number of consultants, and each representative has control over a contiguous sales region.

This problem may be solved using a form of self-organizing map in which the location of each representative is analogous to the neurons, and a 'penalty function' associated with each representative that reflects the proximity of the representative to the customers equates to the variable weights. As above, the algorithm has eight steps:

1. Determine the target workload for each representative. In this case, we have 337 consultants to share between 8 representatives, with a target of 42 consultants per representative. Assign a penalty of 1 to each representative.
2. Find locations for each sales person at random.
3. Examine outliers. In territory management applications, this usually means that care must be exercised at both ends of the spectrum—in places such as Northern Scotland we must ensure that remote areas with relatively few customers do not exercise undue influence; in places such as London, we have to be careful not to overload the representatives.
4. Allocate each KCC to the nearest sales person. Relocate each sales person to minimize the distance travelled to the assigned KCCs.
5. Calculate workloads for each sales person. Increase the penalty values for representatives whose workloads exceed the target, otherwise decrease the penalties. Note that this is a shortcut against the version proposed by Openshaw and Openshaw (1997, see the preceding text), who imply that this step should be applied to a single representative at each iteration. The effects of this change have not been examined.
6. Apply a small reduction in the increments applied at Step 5.
7. Weight each distance from a KCC to a sales point according to the appropriate penalty. Repeat steps 3 to 6 until a stable solution is reached.
8. Map the resulting territories.

An optimized solution to the Spiridon problem is shown in Figure 6.6. We can see that the algorithm is able to generate a very high quality solution to the problem in which workloads are evenly matched within sensible sales territories. Nevertheless, it is also fair to say that such a solution is unlikely to provide the last word within an applied planning context. One possible reason is that, on reflection, it may be supposed that the workloads of each representative are actually far from balanced. Compare, for example, the working week of the representative in Scotland with his colleague in South-East London. It might be fairer to include some element of travelling time within the definition of workloads. However, if one supposes that meetings with consultants are the key issue and travel is relatively unimportant, then this may not be an issue. However, there are many other potential complications:

1. The ideal location for the south-west area representative is Bristol. However, the best candidate lives 50 miles away in Taunton.
2. Birmingham and Coventry are split between different representatives, but they are actually controlled by the same hospital trust. Consistency of message between consultants within the same trust is of the essence.
3. A new cancer centre is under consideration in Carlisle.

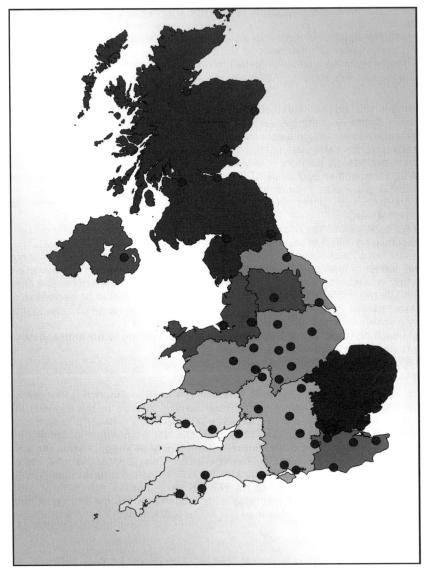

Figure 6.6 Optimized solution to the 'Spiridon' problem

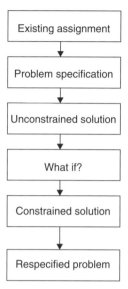

Figure 6.7 Application development cycle for the 'balanced workload' problem

4. The distribution of consultants does not adequately reflect the levels of morbidity for cancer within each region. When the underlying pattern of each disease is considered, then a very different pattern of workloads emerges.

A typical project will require that issues such as these are identified and resolved within the application cycle. A schematic representation of this process might take the form shown in Figure 6.7.

The resultant solutions at any stage of the process may be manipulated using the territory management procedures described in Section 6.2 (i.e. manual reassignments may be made and assessed; benchmark reports may be produced, and so on).

The problem that we have described above has some similarities to Location-Allocation models, which have been of interest of regional scientists and geographers for many years (e.g. Cooper, 1972; Rushton *et al.* 1972; Ghosh and Harche, 1993). However, these authors are unaware of any general-purpose algorithms that could solve the problem specified above. Some more associations between location-allocation models and retail network planning are identified in Chapter 9.

We can now revert to the Paper Planet question with which we began this section. The question can be addressed using an algorithm that is very similar to that outlined above, except that the number of franchise areas (or representatives) is now also a variable to be determined. In practice, this can be tackled by 'embedding' the whole algorithm within a further loop, in which the number of areas is initially fixed arbitrarily. If a viable solution is found, then the number of areas is incremented by one (or vice versa), and so the process continues, until no further valid solutions can be found.

In our discussion of the Spiridon problem, we have seen that achieving balance with respect to a single workload criterion is often a significant oversimplification of the issue (in this case, there may be a balance to be achieved between meeting time and travel time). In the next section, we will move on to consider a range of more complex territory management problems.

6.4   Efficient Organization

Following on from its latest research and clinical trials, Spiridon is now ready to launch a new, more mass-market product that alleviates the problem of high blood pressure among the elderly. In order to promote its product to General Practitioners, the company has elected to focus on Primary Care Groups (PCGs). PCGs are GP collectives that have been established for the purpose of shared protocols on prescribing, referrals, and other clinical issues. There are approximately 500 PCGs in the United Kingdom.

Spiridon's objective is to establish meetings with the Prescribing Leader within each PCG on a one month (20 working day) cycle. Meetings with the Prescribing Leader will be allocated 60 minutes. The intention is to appoint a number of sales representatives who are able to meet this requirement, by travelling between PCGs and a base location on a regular basis. Each sales representative will work from 8 A.M. until 6 P.M., with a one hour break for lunch, travelling by car to the various meetings. The problem is to determine the number and base locations of sales representatives that can meet this workload.

It can be seen that the complexity of this problem is of a different order of magnitude to the questions addressed in the previous sections. The following dimensions all need to be addressed:

- How many representatives should there be?
- Where should the base locations be placed?
- Which representatives should be responsible for which PCGs?
- Which PCGs should be visited by a representative on the same day?
- In which order should the PCGs be visited?

There are some obvious similarities to the well-known 'travelling salesman problem'(TSP), in which an individual needs to visit a number of towns before returning home, and the problem is to determine the order in which the towns are to be visited. This has long been recognized as one of the hardest problems in Operations Research, because it cannot be tackled effectively using the two most favoured computational approaches: brute force (in which every possible solution is evaluated) or hill-climbing methods (in which a method is adopted that can find progressive improvements from an initial solution). Brute force fails because the number of solutions multiplies exponentially for a significant number of towns (say twenty or more). Hill climbing is unreliable because the problem is non-linear, so the method may arrive at one of many solutions that may be far from the best.

An approach to the travelling salesman problem using 'simulated annealing' (see Section 6.3) has been defined by Press *et al.* (1989). This solution is publically available to readers of the book. It may well be that an ideal solution to the 'Representative Routing Problem' (RRP) would be to embed a simulated annealing procedure within a self-organizing map. The self-organizing map would define the territory to be covered by each representative, while simulated annealing could be used to determine the ideal route assignment. It may also be that the 'RRP' is amenable to solution by a 'genetic algorithm' (GA), which has already been shown to be effective to certain highly complex network planning problems (see Chapter 9). At the time of writing, however, the authors have yet to define a GA approach to the much simpler territory planning problem of Section 6.2, so any solution to the RRP remains some way off.

In this section, therefore, we have devised a solution to the RRP involving the recursive use of self-organizing maps. The territories for each representative are derived using the procedure at the top level; then the individual customers are grouped together using a similar procedure. The ideal route through each customer group is determined using 'brute force', that is, every possible combination is evaluated. This is possible because the number of customers who can be visited each day is clearly restricted, both by the time that is needed to reach them and the meeting time that must be allocated to each customer. The data required for the problem is once again the location of each PCG and the drive time between each. In seeking the ideal number of representatives, our strategy is similar to that which was adopted in Section 6.3–in effect, we solve the problem for different numbers and then pick the best.

For a given number of representatives, the solution strategy follows the steps described in the flow diagram of Figure 6.7. The objective is to find a 'balanced distribution' of workloads, which is effectively the length of the cycle in which each representative can

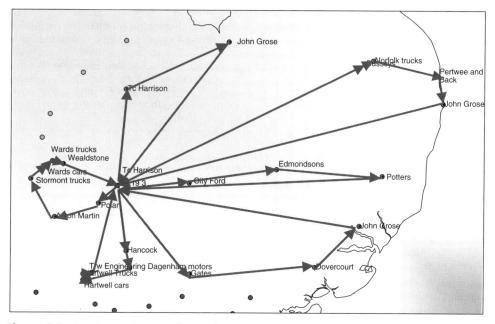

Figure 6.8 Routing assignment for a sales representative

visit all of the PCGs. The key steps in the algorithm are a 'route assignment' module, that allows PCGs to be grouped into sets that can be visited in a single day; and a penalty function module that forces workloads to be equalized between the different territories. The solution is shown in Figure 6.8.

It is our belief that the RRP algorithm described in this section is of more practical significance than the theoretically interesting TSP, which has been exhaustively analysed within the science of Operational Research. The RRP has application to just about every problem of warehousing, distribution and sales force organization:

- A retailer wishes to establish depots to supply all of its stores within 90 minutes drive time. How many are needed?
- A car dealer wishes to organize the delivery of parts to its dealers so that parts are always available on demand within 24 hours. How can this be done efficiently?
- Mobile telephone cells can serve 5000 simultaneous calls within a 30 mile radius. A maximum rate of failure to access the network of 0.5% is permitted. How many cells are needed and where should they be located?

6.5 Conclusion

This chapter has discussed the problem of territory planning with extensive reference to two imaginary businesses: 'Paper Planet' and 'Spiridon'. The case studies have been used in this way to protect the confidentiality of specific client applications. The examples are, however, based on our extensive territory planning experience with companies in a number of business sectors:

In the pharmaceuticals sector, organizations such as Smithkline Beecham and Aventis have used these methods to create ideal territory structures and to manage 'disruption' from new products. In the petrol sector, a 'Market Area Manager' has been developed for Esso to match strengths of the brand against its competitors, in relation to the demand for fuel and other services (see Birkin, Boden and Williams 2002).

In financial services, containment areas have been used for customers of Barclays Bank, Halifax and Portman Building Society as a basis for strategic representation planning, especially before, during and after major mergers (see also Section 9.4).

Automotive manufacturers such as Ford have used territory planning, for example, to systematically restructure networks (see Section 9.2).

Retailers such as Ikea, Dixons, Oxfam and Asda have used this capability to identify zones for local marketing around stores and for planning stock levels and store formats.

Utilities such as Energis have used territory planning as a means to identify distribution points for new services such as business-to-business telecommunications.

Further examples in the use of territory planning in the context of network design and representation are included in Chapters 8 and 9.

METHODS FOR SITE SELECTION

7

7.1 Introduction

In this chapter, we will look at the ways in which the quality of 'sites' may be evaluated. As usual, the sites in question might be petrol stations, bank branches or ATMs, mobile telephone transmitters, shops or any of a variety of retail service outlets. The objective of the chapter is to explore two major themes – the contrast between inductive and deductive approaches to site evaluation, and the contrast between simple and complex approaches. As most of our 'simple' methodologies do involve some form of spatial analysis, it is useful to make some general observations on the store location process. Despite the increase availability of computers and computer software for spatial analysis, a number of recent papers remind us that gut feeling and intuition still play a fundamental role in the retail planning process (Hernandez 1998, Hernandez and Bennison 2000, Clarke I et al. 2000). Clarke I et al. (2000) suggest that this is because more sophisticated techniques either ignore or underplay the retailer's intuitive judgement in the location decision-making process. To that we might add cost implications of more sophisticated techniques (although these will be a fraction of the marketing budget!), and, perhaps, a fear to engage with such advanced technologies. Also, the purveyors of such techniques have not been as successful in their marketing as they might like, with too few examples of the business benefits that might accrue through greater levels of sophistication. We hope to rectify this in the remaining chapters!

However, it is important to realize that location decisions will always have an element of intuition and perhaps cultural traits. Alexander A. (1997) provides a good illustration of the cultural factor in studying the development of Burton's menswear stores in the United Kingdom. As with many other organizations, the choice of individual locations are often made on the whims of senior managers or the Chairman himself/herself. The opening of an early Burton's store in Chesterfield apparently had little to do with a strategic assessment of opportunity. This was, more simply, a temporary home of the Chairman who felt that the town should have one of his stores. The history of store development is littered with such examples. Ray Kroc is supposed to have chosen new locations for MacDonalds stores by flying over them in his plane. It is important, therefore, that we do not berate intuition or cultural backgrounds. Nevertheless, the argument is that for today's retailers there is so much more opportunity to make better informed decisions.

7.2   An Inductive Approach to Site Selection

Let us take as a starting point the sample data shown in Table 7.1. Here, we can see site turnover at a number of locations, together with attribute data about the various locations – the size of the store, the number of people living in the town ('market size'), the number of competing outlets in the town, and the affluence of the population (percentage of households with an income in excess of £50 000 per year). For one location, (Canterbury) the attribute data is known, but not the turnover. Two primary objectives of the evaluation process may be described as follows:

- estimate the turnover or 'site potential' at undeveloped locations (such as Canterbury)
- contrast the effectiveness or 'site performance' of existing locations

In passing, we may note that even our simple example illustrates the different types of data that will typically need to be synthesized in an exercise of this type. Store size and turnover data would normally be supplied directly from a client. The number of competitors might be extracted from a Market Intelligence database such as 'Retail Locations' or BT Business Database. Market size and affluence would usually involve a combination of government sourced demographic data (e.g. census or electoral roll) with either market research ('Target Group Index' TGI, 'Financial Research Survey' FRS[1]) or governmental social research ('Family Expenditure Survey' (FES), 'New Earnings Survey' (NES), 'General Household Survey' (GHS).[2]).

We may also note that each of these headings is essentially a straightforward representation of something that is potentially much more complicated. Thus, 'store size' might be extended to include many different aspects of store quality and 'attractiveness', from the number of parking spaces to layout of the aisles; competition could include elements of brand strength, market overlap and relative pricing; turnover could be segmented by

Table 7.1   Sample store performance data

	Store size	Comps	Market size	% over 50 K	Turnover
Leeds	10 000	7	140 000	25	1 833 000
Manchester	12 000	8	200 000	20	2 747 250
Derby	8 000	4	80 000	20	1 598 400
Sheffield	11 000	5	100 000	15	1 800 750
Croydon	12 000	6	110 000	30	2 108 425
Cambridge	9 000	4	80 000	30	2 025 400
Guildford	10 000	5	120 000	35	2 580 000
Canterbury	8 000	5	100 000	40	?

---

[1] TGI – A market research survey undertaken by the British Market Research Bureau, which monitors spending patterns by different demographic groups.

FRS – A market research survey that looks at product penetration and brand preference for a wide variety of financial products and services.

[2] Research that is published on an annual basis, focusing on various aspects of household composition, behaviour, income, and expenditure.

thousands of individual sales lines; market size could be a representation of different local markets for different products (e.g. a petrol station drawing local customers to its forecourt shop; more diverse customers for its petrol); and affluence could be supplemented by age, lifestage, occupation, ethnicity or other differentiating characteristics. The subtleties of location modelling, and the data that may be used to support it, are explored in more detail in Chapter 8.

Returning to our sample database, albeit simplified in both size and content, we can express the objective of the inductive approach in order to draw out and build upon empirical relationships within the data. All of the data ranks equally, and no prior judgements are made about the relative importance of the different attributes or about their likely effect on performance. Consider first the relationship between store size and turnover that is plotted in Figure 7.1(A). We can see from this distribution a clear but inconsistent pattern of increasing turnover with store size. The trend is illustrated in Figure 7.1(B).

In order to quantify the nature and degree of interrelationship between two variables such as turnover and store size, linear regression is probably the simplest and most popular technique. The regression equation can be computed and displayed as a standard option within Microsoft Excel, as shown in Figure 7.1(C). The two key elements of the analysis are the regression equation and its goodness of fit. The regression equation is a formal representation of the trend-line shown at Figure 7.1(B). In this case, we have $y = 155.78x + 496\,775$, where y is turnover and x is store size. This tells us that every additional square foot of floorspace is expected to generate an additional £156 of revenue (to be precise, £155 and 78 pence!). £500 000 of revenue comes 'for free', whatever the store size (or again more precisely, £496 775). The goodness of fit is shown by an r-squared value, which varies between 0 and 1, where an r-squared of 1 shows a perfect relationship and an r-squared of 0 shows no relationship. A fuller discussion of regression analysis may be found in any good statistics textbook, such as Blalock (1974).

We can repeat the analysis of store size against turnover for each of the different attributes to determine the relationship for each attribute, and its reliability. This information is shown in Figure 7.2 and Table 7.2. In this example, we can see that market size has the best fit to store turnover, with an r-squared of just above 50%. Both competition and affluence are weaker predictors than store size. However, it is the shape of the relationship between competition and turnover that is most worthy of further comment at this point. In common with the other relationships, an increase in the attribute value generates an increase in turnover. In this case, each additional competitor is associated with a turnover uplift of about £145 000. This is counter-intuitive, because our assumption would be that the different outlets in competition for a fixed market and the introduction of new competitors would tend to erode the market share for everyone.

There are clear parallels to well-known relationships of the type shown in Figure 7.3, in which we can see two time series graphs of UK banana imports and Gross Domestic Product (GDP) in the 1950s and 1960s. We can see a very close relationship between the two trend lines. However, there are at least four different conclusions that might be drawn from this analysis:

● that growth in GDP is driven by banana imports

● that growth in banana imports is driven by GDP

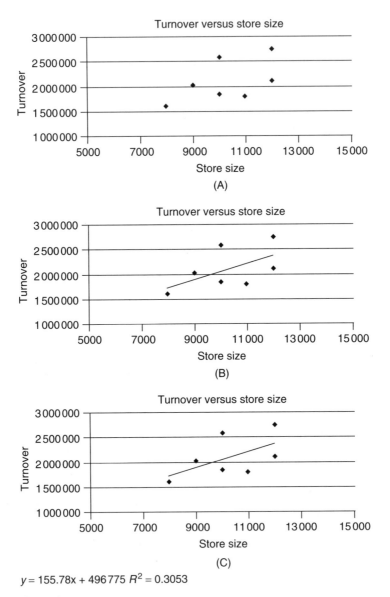

$y = 155.78x + 496\,775\ R^2 = 0.3053$

Figure 7.1 Store size versus turnover

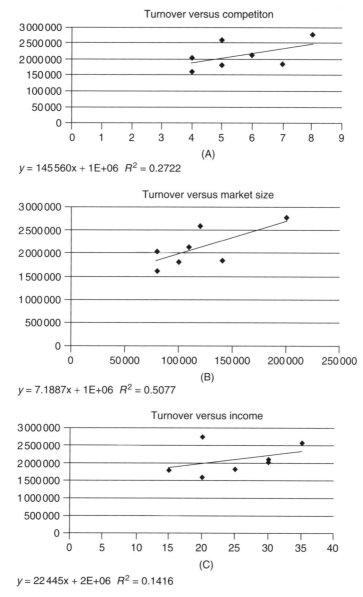

Figure 7.2 Store size versus store attributes

Turnover versus competiton

(A)

$y = 145\,560x + 1\text{E}+06$ $R^2 = 0.2722$

Turnover versus market size

(B)

$y = 7.1887x + 1\text{E}+06$ $R^2 = 0.5077$

Turnover versus income

(C)

$y = 22\,445x + 2\text{E}+06$ $R^2 = 0.1416$

Table 7.2 Analysis of store size versus store attributes

	R-squared	Intercept	Slope
Store size	0.31	496 775	155.78
Comps	0.27	1 288 057	145 560
Market size	0.51	1 246 653	7.19
Income	0.14	1 537 918	22 445

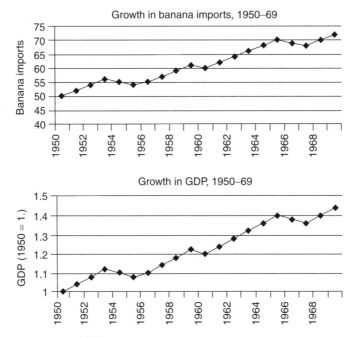

Figure 7.3 Relationship of GDP to banana imports

● that growth in GDP and banana imports are both driven by some unknown but related factor or factors
● that growth in banana imports and GDP is completely coincidental

Of these conclusions, the first is clearly the least plausible, yet, this is almost exactly parallel to an assumption that competition drives turnover. In fact, what we are seeing is something along the lines that large markets provide an opportunity for high turnover, but large markets are also attractive to the greatest number of competitors. This would be similar to conclusion (3) of the bananas example. The problem may be referred to in shorthand as 'intercorrelation', reflecting that each of the attributes is related to one another, in ways that may be complex and difficult to measure.

Once the preliminary analysis of the data is complete, we receive such encouragement from our findings that we wish to build a 'model' that relates turnover to the various store attributes. There are two main ways in which this might be achieved – via multiple regression or site ratings.

Multiple regression is used to combine different attributes into the same regression equation. So in this case, one would have a relationship in the form:

$$Y = a + b_1X_1 + b_2X_2 + b_3X_3 + b_4X_4 \tag{7.1}$$

where y is turnover, X_1 is store size, X_2 is competition, X_3 is market size and X_4 is affluence. a, b_1, b_2, b_3 and b_4 are parameters. The model is usually built up sequentially ('Stepwise'), starting with the most important variable. The compound regression equation is set out at Table 7.3. In this example, we have seen that market size is the most important variable within the individual regressions and this ends up as the most significant variable in the regression equation. Once we have allowed for market size, then the residual effect of competition can be teased out with a more intuitively appealing result. For a detailed discussion of multiple regression, see SPSS (1995).

A ratings approach may be used to provide single scores for each site on the basis of the quality of attributes at that location. Each variable may be weighted according to its relative importance. For example, we have already seen that market size is the most important attribute in our example. This could be weighted at 50, with store size at 30, and affluence and competition both at 10. In this way, a score for each site may be generated between zero and 100, as shown in Table 7.4. In this case, we cannot only obtain a single score summarizing the relative strength of each site, but can also see from where that strength is derived – for instance, Manchester has by far the highest rating, because it also has by far the largest market. Ratings approaches have been most widely applied within fields such as consumer credit scoring (e.g. Leyshon and Thrift (1999)), although the use of 'balanced scorecards' within business planning is another obvious parallel (*e.g. Ahn (2001)*).

As we saw at the outset, two important applications of the model (whether via multiple regression or ratings) are performance assessment and evaluation of potential. On the basis of the ratings, we would most likely identify Guildford and Cambridge as strong performers, with Leeds and Croydon weak, Manchester, Derby and Sheffield roughly at par (e.g. Guildford turnover is well in excess of Croydon, from a lower rating, score).

Table 7.3 Multiple regression model of store performance

	Slope	Contribution to variance
Store size	125	0.20
Comps	116 448	0.18
Market size	5.8	0.33
Income	17 956	0.09
		0.80

Table 7.4 Retail site ratings

	Store size	Comps	Market size	% over 50 K	Turnover
Leeds	20	8	35	63	1 833 000
Manchester	30	2	50	82	2 747 250
Derby	10	10	20	40	1 598 400
Sheffield	25	4	25	54	1 800 750
Croydon	30	6	28	64	2 108 425
Cambridge	15	10	20	45	2 025 400
Guildford	20	4	30	54	2 580 000
Canterbury	10	4	25	39	?

By definition, these variations must arise, either from imperfections in the model data, or from attributes that have not been incorporated. In assessing performance, the user would seek to understand what there is about Croydon that is holding back turnover, or what there is about Guildford that may be reproduced elsewhere. In assessing the potential of Canterbury, the rating may be seen as somewhere between Cambridge (an overperformer) and Sheffield (a slight underperformer). A turnover estimate in the order of £1.9 million would appear to be sensible.

In practice, the authors would tend to favour something like a ratings approach over multiple regression. Although ratings are both simpler and more transparent, it is by no means clear that the method of multiple regression offers greater predictive power. Further discussion on these points follows in Section 7.4.

7.3 A Deductive Approach to Site Selection

Starting from the same data presented in Table 7.1, a deductive approach seeks to develop an understanding of the likely drivers of site performance through intuition or logic. An attempt will then be made to fit this evidence to the model or 'theory'. According to the success of this second stage within the process, it may be necessary to refine, enhance and fundamentally review the theoretical model.

In the case of site selection, we might develop the logic as follows. Turnover at an outlet is likely to result from the combination of two factors – in shorthand, demand and supply. Demand can be measured as the size of the market within a locality. Supply can be represented as the market penetration for the retailer of interest within the local market. In combination, the turnover of an outlet is the product of the size of the market and outlet penetration within that market.

Beginning with market size, we may consider that the available spend is again the product of two factors – the number of people in the area and their average expenditure. For many retailers, that may be the end of the story! Crewe and Lowe (1995) looked at the location strategies of six new fashion retailers in the UK in the early 1990s. For them, there was no perceived need for sophisticated geodemographics or site location strategies. As mentioned in Section 7.1, as with many retailers they went ahead on intuition – seeking out 'affluent centres that are pleasant' (Nottingham, Bath, Oxford, Chester, etc). However, as the number of stores increases, perhaps, the need for more sophistication increases. We can now start to think about concepts such market penetration. Market penetration is, by definition, the ratio of outlet sales to competitor outlet sales. A simple assumption could be that this is approximated by the ratio of outlet floorspace to competitor floorspace. These influences can be brought together to provide a model in the form:

$$D_j^k = \frac{e_j P_j F_j^k}{\sum_k F_j^k} \tag{7.2}$$

where D_j^k represents the turnover of outlet k at location j, P_j is the population at location j and e_j its spending power; F_j^k is the floorspace of outlet k at location j and $\sum_k F_j^k$ is the floorspace of all outlets at location j.

The populations, P_j, and outlet floorspace, F_j^k, are known from the data. These data are repeated in Table 7.5, Columns 1 and 2. The floorspace of each competitor is unknown by location, although we do know the total number of competitors. Let us suppose that each competing outlet has floorspace of 10 000 square feet. The total floorspace for retailer X (excluding Canterbury) is 72 000 square feet and there are 39 competitors. Across all centres, the market share according to the model is 72 000/(390 000 + 72 000) = 15.6%. Market shares by centre are shown in Table 7.5, Column 3.

Store turnover in the seven centres is £15 million. Since this results from a market share of 15.6%, the total market size is estimated at £96 250 000. The population of the seven centres is 830 000, and therefore the average spend per head is £116. However, not all towns are similar in their demographic mix, as we saw earlier. Let us assume that the average spend per head is twenty pounds lower in towns with 25% or less of the population in high income groups; and £20 higher in towns with more than 25% in high income groups. The population of each town is shown in Table 7.5, Column 4, and the average spend in column 5. Multiplying the various factors from the columns of Table 7.5 provides the turnover estimates shown in column 6 of the table.

The question now arises as to how we can assess the effectiveness or 'goodness-of-fit' of our latest model. One way in which this can be done is through the application of a regression equation once more. The turnover estimates of Table 7.5 have an r-squared of 0.81, with a regression line of $y = a + bx$, indicating that the model is effective but

Table 7.5 Retail location data

	Store size	Comps	Market share %	Market size	Average spend £	Expected turnover £	Turnover
Leeds	10 000	7	12.5	140 000	96.0	1 680 000.0	1 833 000
Manchester	12 000	8	13.0	200 000	96.0	2 504 347.8	2 747 250
Derby	8 000	4	16.7	80 000	96.0	1 280 000.0	1 598 400
Sheffield	11 000	5	18.0	100 000	96.0	1 731 147.5	1 800 750
Croydon	12 000	6	16.7	110 000	136.0	2 493 333.3	2 108 425
Cambridge	9 000	4	18.4	80 000	136.0	1 998 367.3	2 025 400
Guildford	10 000	5	16.7	120 000	136.0	2 720 000.0	2 580 000
Canterbury	8 000	5	13.8	100 000	136.0	1 875 862.1	?

that it tends to slightly overpredict the sales of small stores and underpredict the larger ones. An alternative measure is to calculate the Standardized Root-Mean-Squared Error (SRMSE) between the observed and predicted values. SRMSE can be calculated by summing the square of the difference between each turnover and the associated prediction, then calculating the average, and the square root of the average. In this example, we have SRMSE of 225 869. This could be compared with the results of the multiple regression model at Section 7.2, with an SRMSE of 768 340. An excellent review of comparative statistics within location modelling is provided by Knudsen and Fotheringham (1986). Unfortunately, the problem with r-squared, SRMSE, and other statistical measures is that they are not always intuitive to those without prior knowledge or experience. A simpler measure, which is both intuitive and straightforward, is to calculate the average error ('mean absolute error') of the predictions. This tells us that the error margin for the present model is around 9%, which may or may not be significant in a statistical sense, but gives the planner who is trying to use the model a good idea of the reliance that may be placed on it.

From the preceding discussion, we can see that the outputs of our deductive model are not yet quite as reliable as those of the inductive approaches within the previous section. One possible reason for this is that the model has not yet been fully aligned with the appropriate data. To put it another way, the model is only partially calibrated. For example, let us compare the performance of outlets in affluent towns with those of the less affluent, using the previous definitions. We can see that the outlets within affluent towns tend to underperform, and this, in turn, implies that our initial guess at variations in average spend needs to be refined. One reason for this may be the somewhat crude categorization of towns into only two groups of affluence. If a continuous scale is adopted in which cities such as Leeds and Cambridge have moderate levels of affluence, then it is possible to reduce the average errors to around 7%, with an r-squared in excess of 0.9 (see Table 7.6).

We are now in a position to provide turnover estimates for the new store in Canterbury. With an affluent catchment of 100 000 people, the market size estimate is £13.5 million. With a floorspace of 8000 square feet and 5 competitors, the market share estimate is 13.8%. Expected turnover of £1.86 million is implied, which is very similar to the previous ratings estimate. Note, however, that the interpretation of performance in relation to the existing stores, as shown in Table 7.6, is not entirely consistent with our previous assessment. For example, although Croydon still appears to be the weakest performer in the store portfolio, it is Derby that now performs best.

Table 7.6 Store performance data

	Average spend £	Expected turnover	Turnover	Error %	Store performance %
Leeds	116.0	2 030 000	1 833 000	11	90
Manchester	106.0	2 765 217	2 747 250	1	99
Derby	106.0	1 413 333	1 598 400	12	113
Sheffield	96.0	1 731 148	1 800 750	4	104
Croydon	126.0	2 310 000	2 108 425	10	91
Cambridge	126.0	1 851 429	2 025 400	9	109
Guildford	136.0	2 720 000	2 580 000	5	95
Canterbury	136.0	1 875 862	?		

7.4 Applied Location Modelling

In the previous sections, we have looked at two different styles of modelling within the context of a 'toy' example. We now wish to provide an indication of the range and styles of approach that are commonly employed to address 'real' site location problems.

7.4.1 Analogue models

Analogue techniques were (and still are) also very common procedures for site location in the United Kingdom and United States (see Table 7.7). The basic approach involves attempts to forecast the potential sales of a new (or existing) store by drawing comparisons (or analogies) with other stores in the corporate chain that are alike in physical, locational and trade area circumstances. This may be done 'manually' or through regression techniques (see below). Hence, if you are evaluating a new store site in say Cambridge, can you find an existing store location around the United Kingdom that has the same (or similar) population and trading characteristics of Cambridge? If so, you can attempt to draw analogies with the trading performance of the store in that other town. Alternatively, the procedure may work by trying to find sites that are analogous with the top performing stores within the company. That is, if Oxford is performing very strongly, can the analyst find sites elsewhere in the country that match the characteristics of the Oxford site?

In addition to site location, the analogue approach is also suitable for store formatting. Think of a store such as Marks and Spencer, which may have a variety of formats suitable for different locations. Locations that have a big market but not much local competition may be classified as 'urban retail parks'. These will be suitable for megastores with a full product range within a spacious store layout. Locations with a large market, which are

Table 7.7 Use of locational planning techniques (percentage of responding companies)

Technique	Used (% Sample)
Comparative	
Rules of thumb	100
Checklist	63
Analogues	33
Ratio	30
Predictive	
Multiple regression	42
Discriminant analysis	12
Cluster analysis	42
Gravity models	37
Knowledge based	
Expert systems	9
Neural networks	14

Source: 1998 Survey of UK Retailers, (Hernandez, Bennison and Cornelius, 1999)

also competitive, are more likely to be town centre sites, which may be better served by a 'metro' format; perhaps, concentrating on a narrower range of high margin product lines. Small towns with modest competition may provide the environment for a standard layout, for example, a reasonably full range of clothing, and perhaps, foodstuffs, but no bulk product lines such as furniture and houseware. In small towns that are also competitive, it is probably necessary to provide specialized formats such as menswear or food only stores.

Almost a special case of the analogue model is catchment area models based on geodemographic classifications. Geodemographics is essentially an attempt to categorize residential areas according to their similarities. For example, if one takes a single unit postcode, such as LS19 6BA, then this might be classified as an 'ageing professional' location within a geodemographic classification. Other locations, within the city of Leeds or elsewhere, may be found with the same classification. When an outlet or retail centre is profiled according to its market profile, then similar performance levels will tend to be expected from markets with a similar profile.

The success of this approach depends on whether or not you can find similar sites across the country and whether you believe you can successfully transfer the trading characteristics across geographical locations. This again depends on the experience of the location analyst and his/her team. According to Moore and Attewell (1991), p.24, 'Tesco' are improving in this area:

> (The) greater understanding of the way in which existing stores trade has been fed back into the sales-forecasting process through an increased appreciation of analogue store performance.

Apart from the required experience, a second problem remains the variable performance of stores across similar geographical markets. In reality, a wide variation in performance is frequently found between outlets in a retail chain. If a similar geographical catchment is found to the new store, what happens if the analogous store is currently over or underperforming? Thirdly, it is extremely difficult to evaluate green-field sites in this way. These may have catchment areas greatly distorted by local transport networks and it would prove impossible to import revenue predictions from other towns or cities.

A similar approach to the analogue method has been to follow the behaviour of other (larger) retailers and base store location decisions on whatever decisions they make. This has been labelled the *parasitic approach*. In the early days of the British high street, many new multiple groups would simply follow Marks and Spencer, Woolworth and Boots to new locations. The practice is still common today, especially for smaller retailers. In the United States, for example, Mason and Meyer (1981) quote the (then) strategy of County Seat:

> If a Penney, Sears, Wards or local department store is going to go there then they have already done demographic studies. It almost sounds too simple but that really is our strategy.

7.4.2 Geographical information systems (GIS)

Once an estimate has been made concerning the demand within the likely catchment areas (normally used in conjunction with a market survey to see how far people typically travel to a similar store elsewhere in the corporate chain), then a variety of methods may be used to translate population totals into branch sales. The most likely method is the so-called 'fair-share' approach (Beaumont 1991b). Hence, if there are three other competing stores in the buffered catchment area of the new store, then the new store may be expected to obtain 25 percent of the revenue generated in that catchment area. This simple fair-share allocation could be weighted by store size or by retail brand to increase realism. The alternative is to assume the consumer will travel to the nearest store within the catchment area (*dominant store analysis*: see Ireland 1994).

Although the GIS/geodemographic approach is popular with retailers, there are two principal drawbacks. First, there is the problem of how to define the catchment area and second, how to adequately treat the competition. The former is usually represented by distance or drive time bands, and it is often assumed that the store will capture trade uniformly in all directions. Even when drive time bands are drawn in relation to transport networks (Reynolds 1991), there is still the assumption of equal drawing power in all directions. These methods also give equal weight of importance to all households within a buffer. If a five-mile buffer is drawn around a new store as the primary catchment area, then households close to the site are given the same weight (or probability of patronage) as those 4.9 miles away. In addition, the treatment of the competition is wholly inadequate. The presence of competitor stores will mean the real geographical catchment area of a new store will be highly skewed in certain directions. This can normally be shown in all appraisals of existing store catchment areas. Similarly, there is no effective way in most GIS of estimating the new store revenue in light of the level of competition. As we noted above, the method most often used is 'fair share', with the potential revenue of the catchment area being simply divided between all retailers on some ad-hoc basis (type of retailer, level of floorspace, etc.). Hence, this methodology (while offering a useful overview of potential catchment area revenue) is fundamentally flawed because of the inadequate treatment of spatial interactions and the inadequate treatment of competitor impacts (for more details, see Benoit and Clarke 1997).

It should be noted that in more manual catchment area analysis some of these problems have been solved quite effectively (see Davies and Rogers 1984). However, little of this work has so far appeared in GIS packages to sophisticate the level of analysis.

7.4.3 Ratings models

The ratings approach is particularly suitable in highly complex markets, where more sophisticated modelling endeavours are liable to break down. A good example is automated teller machines (ATMs), where there are many outlets, which are frequently but irregularly used by a variety of customers (shoppers, workers, residents, enjoyers of leisure activities such as theatre goers), at a variety of locations (supermarkets, petrol stations, high streets, bank branches) and where a number of alternative distribution channels are available (branches, telephone, internet, supermarket cashback, credit card). GMAP has developed an 'ATM Ratings' product, which uses demographics, retail and business activity, workplace data, competitor activity, and other variables to derive scores for ATM location. ATM Ratings are available as an 'off-the-shelf' standard, but are usually bespoked through the introduction of customer-specific data. In a complex and unpredictable market, these ratings are highly correlated with ATM transaction levels, as illustrated in Figure 7.4. The ratings may be used to score individual sites, or as the basis for 'hotspot' analysis to identify areas in which to concentrate the search for new sites.

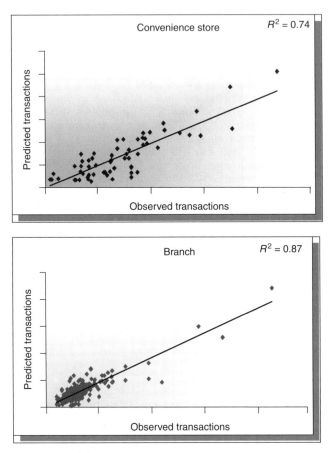

Figure 7.4 ATM ratings model results

An example of a site rating system is shown in Figure 7.5. The system has been developed on behalf of a mixed goods retailer, who typically operates from a large edge-of-town site with good accessibility and car parking. Often, these developments may take place at retail parks with other shops and service outlets (such as cinemas or quick service restaurants). The rating system comprises four components – market size, demographic mix, retail synergies and the supply ratio. The illustration shows a site report for a store to be located quite close to the centre of Leeds.

Market size refers to the number of customers within a fifteen-minute drive time of the store. The market size is weighted in such a way that the number of customers within the first drive time band (i.e. within five minutes) has more importance than the number of customers between 5 and 10 minutes away, and so on. In some applications, these weightings may also be configured to vary according to the location type of the store. For

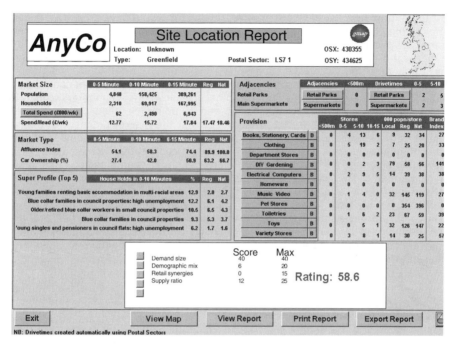

Figure 7.5 Asset rating system

example, population in the first five minutes may be more important at an urban location than at a semi-rural or edge-of-town site. The average spend per head also influences market size, so that we can arrive at an estimate of the total spend as a function of affluence in the local market. The demographic mix of the local area is incorporated within this application via the use of Superprofiles[3] neighbourhood types. In this example, we can see that the major neighbourhood types are young families, blue-collar families and retired workers – in fact, a fairly typical inner city demographic profile. Demographic mix affects the model through both the expenditure profile[4] and brand preferences of the customer base[5].

The 'retail synergies' component measures the effect of other retail activity on the same site. For example, the existence of a major department store at the same location may be

---

[3] Superprofiles is a geodemographic package that is marketed by CDMS. For more discussion of geodemographics, see Chapter 6, Section 2 and Chapter 10, Section 2.

[4] Basic expenditure patterns by affluence are incorporated within the 'market size' component. The subtleties in this process are incorporated through the demographic mix. Thus, young people may spend more on clothing than mature customers, even though they have a similar level of affluence within the market size calculation.

[5] For example, in the UK clothing sector, younger people may have a preference for Next or Benetton; mature customers may prefer Austin Reed or Marks & Spencer.

a substantial draw for customers. Therefore, this factor would have a positive impact on ratings. On the other hand, the presence of other retailers of the same type will tend to compete away the available spend at this location, and this feature is incorporated through the 'supply ratio'. Thus, a large number of competing brands will give rise to a low value for the supply ratio and vice versa.

In the example presented, we can see that demand size is the dominant component and accounts for 40 rating points. A further 20 points are allowed for demographic mix. Retail synergies and the supply ratio account for 15 and 25 points, respectively, making a total of 100. In the example of Figure 7.5, we can see that the market potential of our site is quite high, with a combined score of 46 across the two demand components (i.e. 'demand size' and 'demographic mix'). However, this is quite a competitive location ('supply ratio') with few supplementary attractions ('retail synergies'), thus, a combined score of 12 on the supply components. This yields a moderately attractive combined rating score of 58. The potential value from a rating model of this type is discussed in more detail in Chapter 11.

In effect, what we have here, therefore, is a 'scorecard' of a type that has been widely used in credit-scoring applications by retailers and financial service organizations and recently popularized by management consultancies for business process evaluation and re-engineering (e.g. Talluri (2000)). The application of these techniques to evaluate retail sites has not received widespread consideration in the literature, however.

7.4.4 Regression

Regression modelling is a simple and robust technique for location assessment. In addition to the basic technique of linear multiple regression that was introduced earlier, enhancements that allow the attributes to be transformed in various ways, such as logistic regression, are also possible. Examples of regression modelling applied to retail location problems are provided by Simkin (1990). More recent references are difficult to find, partly because the application of such standard techniques is of little theoretical interest; but also, one suspects, because these have been found wanting in practice. Regression models have been used for single store analysis (a good illustration is provided recently by Morgan and Chintagunta 1997) and to assess entire store networks, usually with more sophisticated multiple regression models. The multiple regression model builds on the philosophy of the analogue procedure (see Rogers and Green 1979). Regression analysis works by defining a dependent variable, such as store turnover and attempting to correlate this with a set of independent or explanatory variables. Coefficients are calculated to weight the importance of each independent variable in explaining the variation in the set of dependent variables. The model can be written as:

$$Y_i = a + b_1 X_{1i} + b_2 X_{2i} + b_3 X_{3i} + \ldots + b_m X_{mi} \tag{7.3}$$

where Y_i is turnover (the dependent variable) of store i
   X_{mi} are independent variables
   b_m are regression coefficients estimated by calibrating against existing stores
   a is the intercept term

Fenwick (1978) gives an example of these variables for a building society. Keeping the above terminology:

X_{1i} = average age of persons in catchment area of branch i

X_{2i} = average socio-economic status in catchment area of branch i

X_{3i} = number of years branch i has been established

X_{4i} = number of new houses under construction in catchment area of branch i

X_{5i} = total number of building societies in the catchment area of branch i

Although these models allow greater sophistication and objectivity than more manual analogue techniques, there remain a number of problems. The primary weakness of such models is that they evaluate sites in isolation, without considering the full impacts of the competition or the company's own global network. As the above building society example shows, the level of competition is typically incorporated by the simple absence or presence of stores. A second major weakness is the problem of 'heterogeneity of sample stores'. This was also seen as a problem with analogue techniques. That is, how easy is it to find a sample of stores that has similar trading characteristics and catchment areas (see Ghosh and McLafferty, 1987)?

A third problem relates to the basic feature of regression analysis, which assumes that the explanatory variables in the models (X_{mi}) be independent of each other and uncorrelated. In many retail applications, this is not the case – independent variables such as floorspace and car parking spaces may be strongly correlated. This can lead to unreliable parameter estimates and severe problems of interpretation. The so-called multi-collinearity problem has received much attention in the literature (Lord and Lynds 1981, Ghosh and McLafferty 1987). However, through careful analysis and interpretation many of these problems can be overcome. Most poor applications of multiple regression in retail analysis have shown statistical naiveté and limited understanding of retail process.

Fourthly, and from our point of view the most important limitation, is that regression models fail to handle adequately *spatial interactions or customer flows*. That is, they do not model the processes (spatial interactions) that generate the flows of revenue between residential or workplace areas and retail outlets. Although regression models may sometimes demonstrate impressive descriptive powers (through their ability to reproduce the variation in sales across a network), the absence of any process modelling leaves us sceptical as to their ability to undertake *impact analysis* with any confidence.

Advocates of artificial intelligence (AI) and high performance computing have sought to breathe new life into statistical modelling techniques, of which regression is the standard. Neural networks became popular in the early 1990s. They offered a predictive modelling approach based on a set of highly non-linear relationships between a dependent variable (e.g. site turnover) and a series of independent variables (e.g. demographics, competition, site attributes). The justification was typically along the lines shown in Figure 7.6. Performance of outlets within a network, or over time, was typically somewhat erratic and complex (Figure 7.6A). Regression approaches would tend to average out these complexities (Figure 7.6B) and when extrapolated would produce an unrealistic trend into the future (Figure 7.6C). The neural network model would be able to handle this complexity

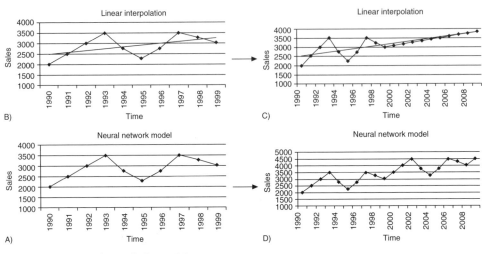

Figure 7.6 Neural nets (schematic)

and, therefore, produce much more realistic forecasts of the future (Figure 7.6D). The weaknesses of the neural network approach are twofold. In the first place, the models are black box, with no obvious internal logic. Without being able to understand variations in historic performance, how is it possible to have confidence in future predictions? And secondly, if the models cannot be understood, how can they be defended against the questions of colleagues or superiors, and if they go wrong, how can they be fixed and (more important) how does the blame get allocated? The authors are not aware of any blue chip retail businesses that are currently using this technology for retail planning.

Other techniques from AI, such as genetic programming, genetic scorecards, or even cellular automata, may also have potential for retail planning – certainly, if one were to believe the words of Openshaw and Openshaw (1997), who provide an excellent review which is, however, noticeably lacking in practical examples. One of the few examples of the application of AI techniques is provided by Birkin *et al.* (2002). These authors discuss the development of a site rating capability for petrol stations. The example is similar in style to the retail application described in Section 7.4.3, although somewhat more complex. The rating for a petrol station might include two components – a location rating, which shows the potential of a piece of real estate to be developed for a specific purpose, for example, a petrol station; and a facility rating, which provides a measure of the effectiveness once a site has been developed. The components of the site rating include factors such as traffic volumes, access and visibility. The components of the facility rating include features such as the size of the forecourt shop, the product range and local price competitiveness. Birkin *et al.* argue that because of the desire for transparency and simplicity (see also Section 7.5) the weights assigned to each component may be sub-optimal. These values may be optimized using a procedure such as a 'genetic scorecard' – essentially a genetic algorithm applied to a scorecard. An illustration of this concept applied to a retail outlet is shown in Table 7.8. The weights derived

Table 7.8 Shop model ratings

Location	Score		Facility	Score	
	Version 1	Optimized		Version 1	Optimized
Traffic	10	8.42	Facia type	5	8.19
Plot area	5	3.61	Development Year	3	4.68
Access	3	2.08	Opening Hours	5	8.32
Market size	20	16.17	floorspace	10	12.65
Visibility	2	1.74	Adjacent retail	5	5.04
Demographic mix	10	7.79	Store format	3	3.95
			Product range	8	7.16
			Competition	4	6.82
			Price index	7	3.38
Total	50	39.81		50	60.19

in association with both location variables and facility variables from a conventional application are contrasted with an optimal set of weights produced by a genetic algorithm. The optimized weights are characterized by more detailed and exact weightings, which are not necessarily helpful, but also by a different balance of emphasis. It may be possible to 'sell' the revised weightings to users of the system on the basis of the enhanced explanatory and predictive power of the optimized model, or simply on the basis of its sophisticated technology. At worst, the results could be used to re-evaluate the composition of the existing model. For example, in this case it appears that down-weighting the importance of the location variables relative to the facility variables would probably be a good idea.

7.4.5 Spatial interaction models

In Section 7.3, we developed a model on the basis of two sets of factors, relating to retail expenditure (demographics, spending habits) and retail supply (store characteristics, competition). The missing element from this mix is accessibility, because as we saw in Chapter 6 previously, customers in Leeds are unlikely to travel to Aberdeen for their weekly grocery shopping. The well-known 'gravity model' of retail trade links the three factors of expenditure, supply and accessibility. This is also one of the oldest retail models, generally attributed to Huff (1963). The model derives its name from the analogue to the physical model of gravitational attraction, which states that the force between two bodies (e.g. planets: analogue – a retailer and a group of consumers) is proportional to the product of the mass of the two bodies (analogue – retail spending and retail floorspace) and inversely proportional to the distance between them. The model has even earlier antecedents in the work of Ravenstein (1885) and Reilly (1931).

The gravity model is one of the most widely used technique in applied retail planning. Versions of the model are offered by all of the major market analysis companies in the UK and elsewhere. Typically, the models use detailed demographics and market research data to evaluate market size and detailed information about competitor characteristics and store performance data on the supply-side. Intelligence relating to demand-supply relationships, such as point-to-point drive times and customer activity patterns, can be used to drive the accessibility component. The integration of diverse data sources, together with substantial enhancements to the model structure, means that the term 'gravity model' does scant justice to the sophistication of the techniques now in use, although the name does still remain popular. We prefer the term of 'spatial interaction model', following Wilson (1967, 1974), to whom the most important model enhancements are due. Spatial interaction models of retail trade are discussed in much more detail in Chapter 8.

7.4.6 Econometric models

We have seen that one of the fundamentals of retail planning is to understand customer behaviour. Econometric models seek to understand the behaviour of individual customers in relation to the choices that are at their disposal. Hence, applications of this class are often referred to as 'discrete choice models', with individual names such as logit, probit, multinomial logit and Dirichlet (see Wrigley, 1988, and Oppewal and Timmermans 2001 for reviews). The classic application of discrete choice models would be to brand selection, for example, what is the propensity to purchase Heinz baked beans rather than another brand in relation to price, the strength of the brand, availability and so on. Such models have also been widely applied to problems of modal choice in transportation research. It has been demonstrated that under appropriate assumptions about the attractiveness of individual retail locations, the econometric and spatial interaction models can become equivalent (e.g. Williams, 1981). One of the main differences tends to be in the form of the variables chosen. Although spatial interaction models can involve highly disaggregated variables (cf Wilson 1983), they are more likely to use a robust proxy for the behaviour they are modelling. With econometric models, a major task is to identify the variables in terms of those that maximize individual consumer utilities. Thus, most studies begin with a major questionnaire or survey to ascertain what factors are most important for each consumer. Bell (1999) provides a good illustration of how many variables may be examined in this way. Inevitably, to operationalize a model, some form of aggregation is needed even with these models.

A good application of an econometric model within a retail planning context would be to analyze prices across a network in relation to local market factors, such as competitor pricing and provision. For example, in the retail petroleum industry, profitability is hugely sensitive to small variations in pump prices, yet the major oil companies still use crude 'me too' rules for pricing in local markets that usually involve a individual dealer matching the price of certain benchmark competitors within the local area. If one assumes that the sensitivity of customers to fuel price can be modelled, then it is possible to produce price recommendations for individual outlets to meet appropriate objectives, such as profit maximization or market share improvement. An example is shown in Figure 7.7, which shows that there is a significant degree of unevenness in actual pump prices. However, when local market factors, such as the concentration of customers and competitiveness of the market, are taken into account, a different pattern of 'ideal' prices emerges. When the variations between actual and ideal prices are considered, there are variations in both directions. Typically, prices tend to run too high in rural areas, where more aggressive pricing strategies could attract more customers and better profits; whereas in urban areas companies could afford to keep prices a little higher when better margins would more than compensate for any dilution of trade.

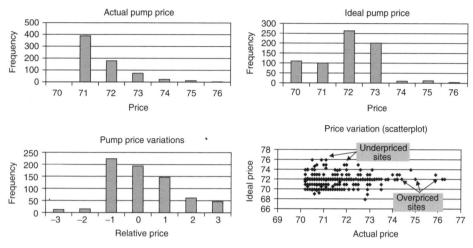

Figure 7.7 Petrol price optimization

7.4.7 Validation of alternative approaches

Whatever modelling approach is adopted for retail planning purposes, at some stage it will be necessary to demonstrate the performance of the model against some objective criterion or criteria. In other words, how good is the model at predicting retail activity or customer behaviour?

In practice, models are most frequently assessed on the basis of historic data. That is, how well does the model explain known variations in performance, perhaps, over time or across a network. This might be quantified using a simple regression equation, and if we have a model with a correlation coefficient (r-squared) of 0.9, then this is better than a different model with a coefficient of 0.8. Analysis of this type is always necessary, but must be treated with caution for the kinds of reasons discussed in 7.2 and 7.4.3. As with financial investments, it is by no means clear that past performance is a good guide to the future, and this is especially the case with models of the inductive style that may lack a clear rationale.

In an ideal world, confidence in a model can be demonstrated when it is able to forecast the impact of change with reliability. This kind of approach is well suited to activities such as direct mail, where the response is usually quite rapid. Thus it is possible to devise a test mailing, set performance targets, and monitor response levels in short timescales before embarking on a full-scale campaign. In location planning, forecasting of this type is more complicated because of the long timescales involved in planning, development and the build up to mature trading. Not only does this mean that any forecasts may be long forgotten by the time that plans become reality, but also, and more importantly, that significant change may also have taken place, for example, new housebuilding, upgrade or change in ownership of a competitor, new roads or closure of a local railway station. Furthermore, it may be rather too late to discover three years later that a key variable has been omitted from the model architecture!

A popular and effective intermediate solution is to use partial data to build the model and then to make blind predictions in relation to information which is known, but of which the modeller is unaware. This is a highly effective, although, nerve-wracking procedure, which should build confidence that the model is logical and unbiased. For example, GMAP built up a revenue-forecasting model for a leading High Street retailer and implemented the model through a cycle of thirty geographical regions. For every six sales totals provided to GMAP for use in calibration of the model, one was withheld and used as a check on the consistency of the model outputs (see Birkin, 1994). Further discussion of the validation of model outputs is provided in Chapter 8.

7.5 The Simplicity Spectrum

We have seen earlier in this chapter that there are numerous approaches to retail network planning.

In Chapter 6, we also began to explore variations in the amount of detail or 'levels of resolution' that may be captured within the models. It follows that for applied modelling purposes there is a huge spectrum of modelling possibilities, from the simplest regression models to highly complex and disaggregate spatial interaction models.

In this section, we seek to make some observations about the relative merits of simplicity versus complexity in retail planning. As a starting point, consider the four conceptual illustrations of Figure 7.8. These show the nature of the relationships between complexity and robustness, cost, transparency and accuracy.

None of the relationships is difficult to understand. A model with one or two variables is always going to be easier to control than a similar model with ten or twenty variables.

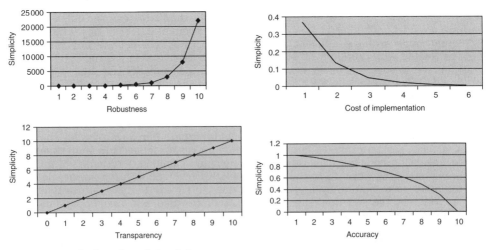

Figure 7.8 The benefits of simplicity

This is not purely a mechanical observation, but also relates to the inherent uncertainties relating to the accuracy of the data, the way it is measured and aggregated and to the way in which the dependent variable(s) are influenced. Thus, a simple model is usually the most robust. More complex models need to be specified in more detail, they need more data to feed them, and the interpretation of their outputs requires greater attention. Once a complex model is established, more training will be required to keep its users well informed. Complex models are, therefore, more costly than simple ones. Retail planning is ultimately a political process in which is not just the results, but the degree of conviction in those results that counts. Thus the level of understanding and transparency within the planning model is an important consideration. The more complex a model becomes, the fewer people who will genuinely understand it, and the more difficult that it will become to explain and share that understanding. Lastly, however, more complex models should almost always be more accurate than simpler ones through the incorporation of more factors, more data and more sophisticated interrelationships.

In bringing together these relationships, it may be seen that the objective of the retail planning process should have an important impact on the level of sophistication that needs to be adopted. Two examples will help consolidate the point.

A multinational oil company wishes to implement a site-assessment tool that quantifies potential fuel sales at different locations. The tool is needed to allocate investment priorities between different countries, as well as between areas and sites within a country. It is therefore necessary to adopt a robust approach using comparable data that is universally available in different places; and which can be understood and interpreted by users within many different countries. Since the tool will also provide a basis for the allocation of resources, it will be important that the outputs are not only fair but are seen to be fair. A relatively simple modelling approach such as a site ratings methodology is highly appropriate for this problem.

A DIY retailer has experienced persistent problems in trying to develop its network through new store openings. It is unable to gauge accurately the potential business levels from place to place. This means that it is unable to provide stores of the appropriate size and format; many stores turn out to be unprofitable; and many opportunities for new development are missed because of uncertainty about the likely returns from the investment. The retailer needs a tool that can predict the impact of a new store opening (or change in format, or other change) with the greatest possible accuracy. The tool will be judged by its effectiveness rather than the clarity of its internal mechanics. A highly disaggregate spatial interaction model will, almost certainly, be the most appropriate choice in this situation.

In summary, therefore, it can be seen that increasing complexity brings greater model accuracy, but only at the expense of a loss of robustness, reduced understanding and increased cost of implementation and maintenance. The fundamental principle of 'Ockham's Razor' has been known for many centuries since its statement by William of Ockham (1280–1347), a medieval philosopher and early Oxford scholar:

> The mind should not multiply entities beyond necessity. What can be done with fewer... is done in vain with more.

In plain English, don't overcomplicate things unnecessarily. Always use something simple, if you can get away with it!

SPATIAL INTERACTION MODELS FOR RETAIL SITE ASSESSMENT

8.1 The Model and its Development

In Chapter 7, we argued that as we move from gut feeling to spatial modelling techniques for store location research, the level of accuracy of the predictions of customer flows and store revenues increases. In this chapter we wish to review one modelling approach in more detail – the spatial interaction model, or as it is still commonly referred to in the retail trade, the gravity model (which is unfortunate since it makes the models sound old-fashioned and outdated when considerable research continues to be undertaken). First, however, it is interesting to reflect again on why this sort of modelling technique is not in widespread use by retail organizations (Hernandez 1998, Hernandez and Bennison 2000, Clarke I et al. 2000), especially as we would expect a good model to be able to reproduce existing flows and revenue totals within 10% of reality. This is a very low error margin in an environment in which individual consumer tastes and behaviours can make quite a difference. We saw, in Chapter 1, that Clarke I et al. (2000) believe that the models do not incorporate the intuitive feelings of key decision makers and that this may be one reason for the lack of applications. Also, Simkin (1990) observes:

> While mathematical models have been created, there is a dearth of operationally predictive models capable of reproducing meaningful and useable information for a company's management.
>
> (p.33)

The point here is that there are few commercial packages available to enable a retail analyst to run a suite of spatial interaction models. The analyst has to do one of three things. First, he/she could program the models themselves. However, this obviously needs good programming skills as well as a good understanding of the properties of the models. Second, the analyst could buy into consultancy services, but many feel this is too expensive. Third, the analyst could use one of the proprietary models available in some geographical information systems (GIS) packages, such as Arc/Info. These are commercial packages and, in theory, ought to have increased the use of such models in the retail industry. However, as we argue in detail elsewhere (Benoit and Clarke 1997), these models tend to be very aggregate (and simple) models and prove to be poor predictors in complicated retail markets. As we shall argue in the following text, real application models usually require a high degree of disaggregation, or customization. As geographers and spatial analysts, we need to convince retailers that there is value in investing in these customized models (by hiring skilled individuals or contracting out). We attempt this below and again in Chapters 9 and 11.

Before looking at such customization, it is useful to repeat the form of the aggregate model. Let us label any residential zone such as a postal sector or enumeration district (i) and any facility location such as a centre or supermarket (j). Then the number of people travelling between i and j can be labeled S_{ij}, where:

$$S_{ij} = A_i \times O_i \times W_j \times f(c_{ij}) \qquad (8.1)$$

and,
S_{ij} is the flow of people or money from residential area i to shopping centre j
O_i is a measure of demand in area i
W_j is a measure of the attractiveness of centre j
c_{ij} is a measure of the cost of travel or distance between i and j

A_i is a balancing factor that takes account of the competition and ensures that all demand is allocated to centres in the region. Formally it is written as:

$$A_i = \frac{1}{\sum_j W_j \times f(c_{ij})} \qquad (8.2)$$

The model thus allocates flows of expenditure between origin and destination zones on the basis of two main hypotheses:

(1) Flows between an origin and destination will be proportional to the relative attractiveness of that destination vis a vis all other competing destinations
(2) Flows between an origin and destination will be proportional to the relative accessibility of that destination vis a vis all other competing destinations

The origin and destination zones have taken many forms in previous studies. In the majority of applications, analysts use the smallest zones possible. In the United Kingdom, this would tend to be postal sectors (communes in the rest of Europe) or enumeration districts (EDs). The latter comprise between 50 and 200 households. The destinations tend to be individual shops (point locations) or shopping centres (zone locations), depending on the application. There are a number of issues of concern regarding the spatial zoning system. The most difficult issue in the past has been finding a study region that minimizes cross-boundary flows, both in and out of the study region. However, because computers are now more powerful, it is possible to apply models at finer levels of detail than previously. For example, in early GMAP projects, it was necessary to carve Britain up into smaller regions, and to model flows between postal districts and retail outlets for a small number of products – typically between 5 and 10. Thus, each model would comprise something like 100 postal districts, 50 retail centres, and 6 products, that is, 30 000 potential flows. Among the practical weaknesses of this strategy is the introduction of 'boundary effects' within the models[1]. In order to cope with these boundary effects, the regions were allowed to overlap, but this, in turn, meant that in at least one case a single retail centre (Marlow) was included within four different regions! Finding a consistent calibration so that Marlow attracted the same number of customers in each of its regional incarnations was no mean feat!! Now it is possible to model flows from 150 000 enumeration districts to more than 1000 supermarkets and retail outlets, for hundreds of products, without the influence of boundary effects. In part, this can be achieved because of computational advances in the storage and processing of data; but the capability has also been promoted by the invention and refinement of new methods such as 'boundary-free modelling'[2] and the 'turbo-model'[3].

Having set up the origin and destination zones, the model works on the assumption that in general, when choosing between centres that are equally accessible, shoppers show a preference for the more attractive centre (which can be measured by size or other attributes such as car parking availability, price etc: see Section 8.3). When centres are equally attractive, shoppers show a preference for the more accessible centre. Note, however, that these preferences are not deterministic. Thus, when choosing between equally accessible centres, shoppers do not always choose the most attractive. The models are, therefore, able to represent the stochastic nature of consumer behaviour. Neighbouring households would not be expected to behave in exactly the same way, even though their characteristics are similar. Equally, particular individuals and households will not always use the same retail centres.

These models have been adapted to a wide range of application areas. They have been used to examine flows of people to shops, offices, work, schools, hospitals and even pubs and dry ski slopes! Indeed, Fischer and Getis (1999) remark that models of spatial interaction have been fundamental in regional science. A detailed history of the formulation, development and theoretical insights of these models takes us well beyond the scope of this chapter. Interested readers can follow this development chronologically through key texts such as Wilson (1974), Fotheringham and O'Kelly (1989), Sen and Smith (1995) and Fischer and Getis (1999). We attempt below to summarize some key developments of the models, but concentrate our efforts in the remainder of the chapter on developments largely driven by commercial applications.

First, the retail model has been widely used as a test-bed to explore a wide range of theoretical ideas in urban and regional analysis. These include the extension of the models to include dynamics and alternative methods for testing of alternative methods of determining equilibrium solutions (Harris and Wilson 1978, Clarke M. and Wilson 1983,

---

[1] A boundary effect occurs when a region boundary is drawn close to the perimeter of 'any town', because if the attractiveness of 'any town' is significant, then customers will be drawn from across the region boundary... but these customers are not accounted for in the model, because they are in a different region. Although the reverse is also true of customers who live in the region and shop outside it, there are no guarantees that the two components will balance out.

[2] Boundary-free modelling exploits the fact that shopping opportunities in Yorkshire have little practical impact on customers in Cornwall as a means for building up a modelling approach that is free of edge effects.

[3] The turbo-model exploits similarities in the interaction patterns between neighbouring residential zones to facilitate an order of magnitude increase in the speed at which retail flows may be evaluated.

1985, Phiri 1980, Rijk and Vorst 1983a,b, Fotheringham and Knudsen 1986 and Kantorvich 1992). The most obvious example of 'applied' use was the presentation of future trajectories of various urban structures, given key changes in the parameters of the models. Empirical research largely took the form of a series of 'numerical experiments' concerning the properties of this model and its variables. However, this period of research clearly demonstrated the possibility of rapid structural change from one kind of equilibrium solution to another at critical parameter values of key variables (Wilson 1981). Clarke G. *et al.* (1998) provide a useful summary of progress with theoretical and empirical work on dynamic models.

Second, extensions have been made to the theoretical properties of the models. These include relatively straightforward experiments to find more realistic formulations of the major variables. Some of these will be detailed in Section 8.3. In a few cases, however, new versions of the basic models have been presented. Fotheringham (1983, 1986) for example, experimented with new variables in the model to address the issues of agglomeration and spatial competition. These are particularly important issues if individual stores rather than shopping centres represent the supply side. The agglomeration or *competing destination* variable allows consumer's utility to increase for comparison shopping by selecting a retail outlet in close proximity to other outlets. Others have followed this line of research. Guldmann (1999), for example, develops and tests a model containing both competing destination and intervening opportunity factors (see also Fik and Mulligan 1990 and Roy 1999 for new hybrid models).

The third set of developments are labelled 'technical' by Fischer and Getis (1999). They refer to work on designing new models of interaction based largely on inductive modelling (see Chapter 7 for more on inductive versus deductive modelling). The idea here is to find models that produce perfect fits between model outputs and real interaction data. New tools, such as genetic algorithms and neural nets, have been extensively used to search for these new models. Key works include Openshaw (1993), Diplock (1998), Fischer *et al.* (1999). The difficulty with these models is that they become less reliable for predictive work. That is, although they fit existing data well, the same form of the models may not fit new data very well. The proof that these models work as well as traditional interaction models in predictive work remains to be presented.

In terms of applications, the most important issue with these models is that they need to be disaggregated in order to fit real world data. The history of spatial interaction models is filled with examples of how the basic terms, introduced in Equation 8.1, can be refined or made more sophisticated. We shall look at the basic demand and supply terms in more detail in the following text. The fitting of the models to real-world data is also non-trivial. There are many parameters (or weights) associated with each variable. All of these need to be estimated (or calibrated), using some kind of statistical goodness-of-fit procedure. In other words, the parameters that minimize the difference between the predicted and observed set of flows are the ones that are chosen. However, the calibration procedures of mathematical models are less problematic when good interaction data is available and when the models are not highly disaggregated. Many organizations are becoming 'data-rich' and many will have, at least, some information on customer flows. If not, this is increasingly available from commercial agents or can be obtained using some kind of questionnaire or sample procedure. Two very significant areas of data collection are the automated collection of electronic point-of-sale (EPOS) data and the generation of huge volumes of customer data, for example, from store loyalty cards.

Point-of-sale data allows sales variations to be monitored, not simply from season-to-season but from day-to-day. It allows product sales by outlet to be tracked in very great detail, that is, at the level of individual products such as tins of Heinz baked beans. Customer loyalty data allows retailers to track exactly who is buying what, and where. This provides high-value intelligence, especially when considering the interaction component of the models. It is also worth noting that one of the issues with loyalty data is that it is not complete, only relating to those customers who choose to use loyalty cards. One of the great beauties of a modelling approach is that this is not a significant obstacle, as one can simply calibrate against that sample of customers who are represented.

In the rest of the chapter, we look at the key terms in the model in more detail and also give some examples of how the models can be used. To set the scene, it is useful to introduce some key model outputs. The text here follows Clarke and Clarke (2001). First, to recap, the model is calibrated to reproduce existing interaction patterns between populations (either at home or at work) and shopping centres. This facilitates the estimation of store turnovers (which may not be available from published sources: few retailers have turnover estimates for their competitors for example). Having allocated expenditures between all retailers in this way, the most obvious new geographical indicator is that of local market penetration. Such indicators show that market share can vary enormously within regions. In addition to new information such as local market shares, the models can also be used to compare actual turnovers to model predictions that is, given a certain population size, type and nature of the distribution of all competitor outlets, what would the model expect a certain outlet to be achieve in sales terms? This helps to provide a more objective picture of store potential. Is a store that earns more than $3 million per annum doing well or badly in relative rather than absolute terms?

The models are most often used in what-if? fashion. Having identified the variations in market penetration, the retailer may be keen to improve its performance by opening new outlets in areas that currently have a low market share. The models can then be used to test the impact of a new store opening. The results for a new car dealership in Blackpool, Lancashire, United Kingdom, are shown in Table 8.1. Note here that the retailer is not only interested in the total sales predicted; the impact on the rest of his/her network is also crucial. In this case, the new Blackpool store generates $300 000 of new business that is mostly taken from the competition.

This is the core output from any modelling exercise. However, we shall develop more outputs and performance indicators below. First, however, it is useful to explore the main variables in the model in more detail.

Table 8.1 Total sales in Blackpool ($000)

New outlet sales = 384		Net company gain = 300	
Deflections from existing outlets			
Lytham St Annes	21	Fleetwood	18
Fylde	20	Morecambe	5
Preston	15	Lancaster	5

8.2 Estimating the Small Area Demands for Products and Services

Measuring small area demand for products or services serves a number of useful purposes for retailers. First, it allows a quantification of available expenditure by small area, which can be aggregated to produce total demand estimates for store catchments or regions (see Chapter 6). Estimating demand at a small area level usually requires combining demographic lifestyle market research with client data. Some markets are easier than others, particularly where organizations collectively decided to syndicate their sales or registration data to produce full market size estimates. For example, in most European countries, car manufacturers pull their vehicle registration data by small area and this historical data can be used as proxy for future demand. Figure 8.1 maps out the historical car registrations for the city of Madrid in Spain for 1997 by commune. However, in the car market (because of the complexities generated by the fleet sector), it is often the case for these historical registrations to contain distortions in areas where there are large fleet operators, manufacturers' headquarters, or daily rental companies. To adjust the registration statistics to reflect the 'true' retail market requires the elimination of these distortions. Figure 8.2 shows the adjustments that had to be made to the Madrid data

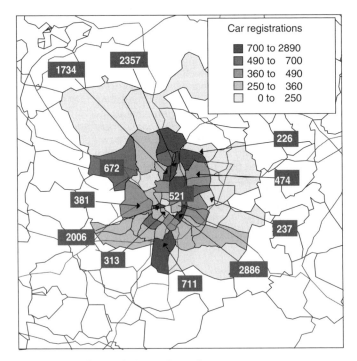

Figure 8.1 Registrations for Madrid (unadjusted)

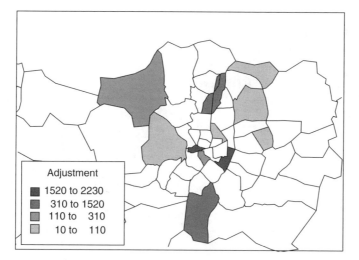

Figure 8.2 Adjusted registrations for Madrid

to create a more realistic picture of available demand. Thus, even with data on observed shopping habits, there can be errors and distortions to be aware of. Other sectors where sharing of sales data is commonplace include the retail financial service industry. For example, in the United Kingdom, CACI operate a syndicated market sized service for financial service companies, covering both mortgage and savings products. Most UK financial services organizations participate in this service, supplying CACI with their sales data by postal sector. CACI then acts as a broker to produce market size estimates for each postal sector, which are then supplied back to the participating organizations. Figure 8.3 provides an example of this data, showing total market size for new mortgages, which are generated through a direct channel (i.e. branch) for a part of south east England. Alongside this market size is an estimate of a particular UK bank's share of this business.

Where no direct data are available, some form of estimation of market size is required. Most particularly, this involves applying market research data directly or indirectly to small area population data to derive demand estimates. Many companies collect data on the social class and age profiles of their customers, either through market research or through capturing point-of-sale data. In every European country, census data is available at the small area level, which will easily yield similar population profiles for small areas. The two sets of data can then be combined to yield demand estimates. For example, if we were attempting to generate estimates of food expenditure by small area in the United Kingdom, the following approach could be used. From the Family Expenditure Survey(FES), it is possible to derive the average weekly expenditure by households in different social class groups (A, B, C1, C2, D, E: the so-called Jictnar classification). From the 1991 census data sets, we can obtain the number of households by social class for each postal sector. A simple multiplication of average weekly expenditure by population total for each social class group will then generate an estimate of available demand. Figure 8.4 shows the

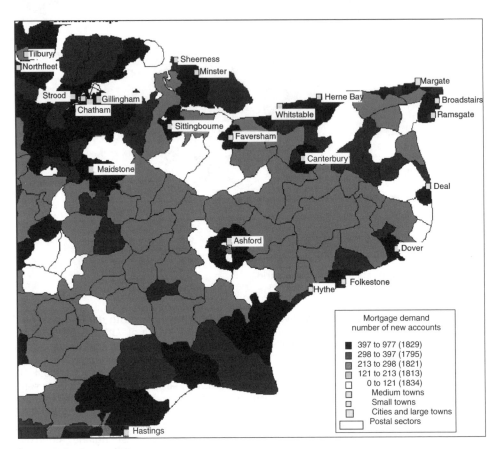

Figure 8.3 Demand for mortgages

results of this exercise for the North West of England and, in this case, we have overlaid the locations of the principal supermarkets in this region. It is obviously possible to refine this analysis to take account of other factors such as household size, number of children and so on, but the principles remain the same.

By producing estimates at the smallest areas for which data are available, it is possible to aggregate the demand estimates to produce information for entire cities or regions. The methodology to create the maps shown in Figure 5.5 is the same as that for the grocery example given in the preceding text. Estimates of the usage of different types of car parts are obtained for different population types. Then, the number of each of these different population groups is obtained from the small area Census of Population. Multiplying these population counts by the average spend associated with each population group produces the required estimates.

Demand - the basics...

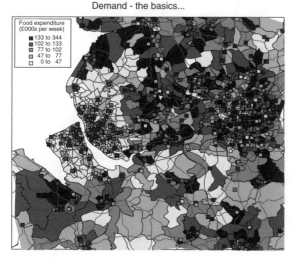

- Estimate demand at small area level

- Combine demographic, lifestyle, market research and client data

- Some markets are easier than others - e.g. any where customer registration is required

Figure 8.4 Demand estimates for UK food expenditure

In some markets, the demand for goods or services might be related to the availability of supply outlets that provide that product or service. For example, Plate 1 shows the average number of ATM transactions per account for postal sectors within north and west Yorkshire. We also overlay on this map the location of branches without ATMs, branches with ATMs and remote ATMs. It can be clearly seen from this figure that ATM usage varies considerably across the region, but is strongly influenced by the availability of ATM machines. Other sectors that show significant amounts of elasticity of demand include recreation, leisure and sport.

Most of these demand estimates are thus proxies, even in the data-rich financial service and car markets. There are, thus, a few problems with demand estimates to be aware of. First, the methods mentioned above make use of broad social class or occupation categories to work out zonal expenditure totals. These categories tend to underestimate the extremes of the population. For example, in the most affluent areas of cities, many residents will be simply classed as social class A/B or 'professional'. There can be a wide variation in each of these categories. 'Professional', for example, could include both teachers and managing directors. Thus, in the very highest income areas, which include far more of the latter than the former, expenditure could be seriously underestimated. Ideally, retailers would like more reliable small area income estimates. These are available in some countries, but not all. In the United Kingdom, there is ongoing research to find the best way to calculate small area incomes (see Birkin and Clarke 1989, Green 1998, Bramley and Lancaster 1998). A second important issue with demand estimates relates to the size of zone. In small zones, population can be assumed to be fairly evenly distributed across that zone without any real problem. In large postal sectors or Eds, the problem is where to allocate that demand. Sometimes, the large zones can be artificially split to distribute the population according to known towns or villages within these zones (thereby

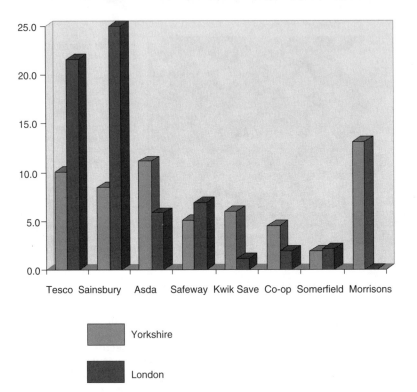

Figure 8.5 Supermarket brand attractiveness

creating extra zones in the study region). If this is not done, the population is normally assigned to the centroid of that zone, which may be misleading in terms of real population distribution. Locating demand at the centroid could also adversely affect the workings of the distance deterrence term, since that demand may have further to travel than demand located within one small part of the zone. To get around this problem, we may wish to use a weighted centroid system, whereby the majority of the population is 'located' at a single point, which is not the centre of the zone.

As with the supply side below, there are many ways of estimating demand, and a number of practical issues relating to those estimates. The reader can start to understand that modelling is as much an art as a science!

8.3 The Supply Side

The next question is, how do we define the attractiveness of individual stores or shopping centres – the W_j term. Traditional applications of the model have assumed that something such as the floorspace of a store provides an ideal proxy for attractiveness. Normally, the bigger the size of store or centre the more attractive it is to consumers, because the bigger size will also normally bring more car parking spaces, more choice of goods and cheaper prices, etc. However, in some cases, it is deemed important to add other variables, in addition to floorspace. First, recognition of different types of consumers, such as car owners and non-car owners, is important in most real-world applications. To non-car owners, the availability of bus services may be important, whereas to car owners the most important variable of attractiveness may be ease of parking. Second, as mentioned above, the destination attractiveness term can be disaggregated to include all sorts of centre or store attributes (Pacione 1974, Spencer 1978, Timmermans 1981, Wilson 1983, Clarke and Clarke 2001). These include the safety and comfort of shopping (e.g. the degree to which an area is pedestrianized and sheltered from the elements), the availability of other important stores, the number of cafés and leisure facilities, and so on. Fotheringham (1983, 1986) has argued that the models need to be modified structurally in order to allow stores in close proximity to other stores to have greater attractiveness to consumers. He introduces a new variable (A_j) into the models to measure the relative accessibility of a destination to all other destinations. His later work incorporates a hierarchical choice component, so that customers are first envisaged as choosing between a cluster of outlets, and subsequently selecting a particular destination from within the chosen cluster (Fotheringham 1986, 1988). The new accessibility term has an associated parameter. When this is greater than one, stores located close to each other have a locational advantage over isolated outlets. This may be particularly important in comparison-shopping. When the parameter is less than one, there is no benefit of being close to other stores (perhaps best reflecting convenience retailing).

Another key term in an attractiveness variable may be the retail fascia, or the name above the door. It is clear, for example, that certain retailers have a more popular image or brand awareness. Thus, it is common now to see retail fascia added to the attractiveness term in the models. This is especially important, as spatially that brand image may be very different across different parts of regions or, indeed, cities. Figure 8.5 shows variations in the relative attractiveness of 1000 square feet of supermarket space between the major brands for two UK regions. It can be seen that whereas Tesco and Sainsbury's have a dominant influence in the south, they are overtaken by both Morrisons and Asda in the north of England.

In a more local context, Sainsbury's will be more important to customers in more affluent areas than say discount retailers such as Aldi or Netto. The opposite is, of course, true in less affluent areas.

When one considers the history of spatial interaction models, improvements to the specification of the models is, by far, the most significant factor. As we have seen in the preceding text, the attractiveness component of the models can now be constructed to a high level of disaggregation. Traditional floorspace indices need to be qualified by brand attractiveness (how magnetic is 10 000 square feet of Sainsbury's relative to 10 000 square feet of Netto?), by brand positioning (how does Sainsbury's appeal to low income groups compare with Netto), and by brand loyalty (how important is it that people have always used Sainsbury's, and what factors encourage brand switching); by the layout of stores, the quality of their management and their level of maturity; by the proximity of other outlets and whether these outlets are competing or are complementary; and by store attributes such as availability of parking, size and quality of its frontage, ease of access and so on.

So, a large number of possible variables can be included in the supply-side attractiveness term of the model. In some cases, adding extra variables can significantly increase the predictive power of the models (Clarke and Clarke 2001). In other cases, the time and effort required to measure these variables empirically may simply not be worthwhile, as no major gain is obtained in terms of greater accuracy. Once again, it is a matter of being able to examine model results and make decisions regarding whether extra variables would be worthwhile. No hard and fast rules apply, unfortunately. This again makes it very difficult to package these models successfully.

8.4 Customer Flows and Store Revenues

The last major term to be considered in the model is the distance deterrence term. This can also be measured in a number of different ways. This term is important because it invokes what is often called the first rule of geography – the principle of distance decay. That is, consumers are unwilling to travel long distances when other closer opportunities exist. The distance deterrence term has often been measured simply as straight-line distance in the past. Today the availability of digital road networks has enabled some organizations to produce travel time matrices, on the basis of average speeds attached to these networks. Given that these travel times are likely to take account of barriers such as rivers, railway lines, motorways, etc., they should be used in preference to straight-line distance.

Different patterns of customer flows may be represented within the model using variations in the distance deterrence parameter (β in Equation 8.1). This can allow markets with very different characteristics to be modelled within the same framework, for example, Figure 8.6 compares the catchment area of Leeds for two products, books and newspapers. The interaction pattern for books would be represented with a much lower value for β than for newspapers. However, as retail markets tend to become more complex, still more subtlety is necessary, owing to increasing complexity in relation not only to customer behaviour but also to retail provision. Customer behaviour is increasingly complex as a result of multi-purpose trip making, in which the pattern of home-shop-home is increasingly replaced by home-school-work-leisure activity-bank-shop-restaurant-home, or some variant thereof. As we saw in Chapter 3, retail provision is becoming increasingly complex through the provision of different channels, for example, the augmentation of branches and ATMs, first by postal and telephone banking, and then by internet and mobile phone services (see also below, and Chapter 5).

In modelling any retail market, the first thing that needs to be established is the existing pattern of retail activity that can be reproduced rationally and accurately by the model. If the model is well calibrated, then the pattern of customer flows from individual demand areas to individual supply locations will be broadly reproduced, see example, Figure 8.7. If the individual flows are modelled accurately, then the summing at each supply point in order to provide an estimate of store revenue could also expect to be accurate. This is obviously vital for retailers, who will wish to have maximum confidence in any revenue estimates to be provided by the model for new store openings, or indeed refurbishments, relocations or closures. Unfortunately, it is difficult to provide verification of the planning capability of a model before new stores are built. In order to get round this, it is common practice to withhold store performance data during the calibration process. Suppose that 200 stores are to be modelled, and 20 observations are withheld. If the model operates

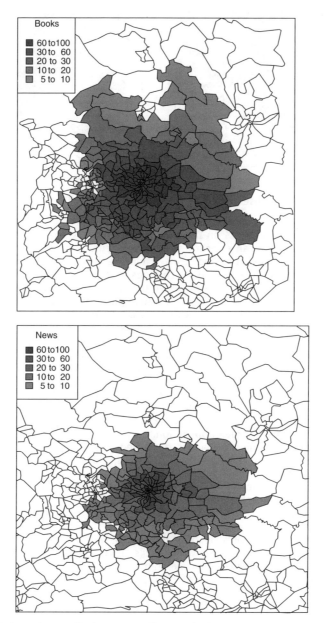

Figure 8.6 Catchment areas for two competing products

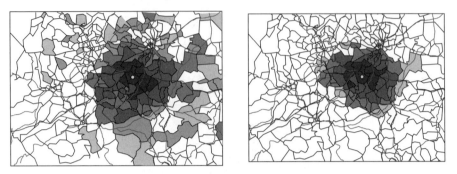

Observed customer flows   Predicted customer flows

Figure 8.7 Retail flows – observed and predicted

to a predictive accuracy of ±10% on the 180 stores it knows about, then as long as the calibration is unbiased, we can expect a predictive accuracy of ±10% on the 20 missing stores. If all of this can be achieved, then it can be further supposed that the same predictive accuracy might be expected for changes to the store network.

Good examples of this process are provided by Birkin *et al.* (1996) and Clarke & Clarke (2001). Birkin *et al.* (1996) showed in a modelling application for a DIY retailer in the south of England, predictive forecasting accuracy on store sales of ±8% could be achieved. Clarke & Clarke (2001) report on the results from a similar exercise applied to modelling fuel volume for petrol stations in Sicily. This is an extremely complex task, given the nature of fuel purchasing. Whereas supermarkets and other large retail outlets are largely dependent on a local catchment population whose extent and purchasing power can be systematically evaluated, there are far more unknowns with fuel retailing. For example, especially on motorways and trunk roads, purchasing by non-local customers ('transient' trade) is much more important. Similarly, the importance of workplaces is highly significant (cf. Birkin *et al.* 2002). Nevertheless, Clarke & Clarke (2001) were able to demonstrate modelling accuracy of ±14% for this complex application.

One of the features of the approaches described in this chapter is that they tend to focus on the sales impact of new outlets or channels. This is very much the hardest part of the forecasting exercise, involving many unknowns in terms of customer behaviour, competitor response, changes in the market environment (e.g. new fashions) and so on. In order to provide a complete picture, however, it is also necessary to look at the cost of new store developments or other network impacts. In order to achieve this an 'investment appraisal' capability is required.

An example of this capability is now illustrated. The example relates to a mixed goods retailer or department store, which sells a variety of goods from clothing to household appliances (housewares). A substantial upgrade to the existing premises is under evaluation, and, using a suite of disaggregate spatial interaction models, sales forecasts have been generated for each of the major product areas. The results are summarized in Figure 8.8. Through its new and improved infrastructure, the store is able to generate enhanced revenues across all of its major product lines, with the biggest uplift in ladieswear. In order to assess the effectiveness of its investment, future sales

Investment appraisal model: summary report

Annual turnover £ 34 115 650

	NPV at 12% £	Average margin %
Menswear	1 752 180	5.43
Ladieswear	3 128 270	10.95
Childrenswear	1 259 379	8.56
Furniture	876 090	7.24
Kitchenware	1 149 868	5.83
Home & garden	1 697 424	9.11
Other	1 587 913	8.07
Total earnings	11 451 124	7.88
Retail overheads	−1 970 996	−5.78
Supply and distribution	−2 211 619	−6.48
Property costs	−2 847 992	−8.35
Third party fees	−1 345 046	−3.94
Credit card costs	− 767 187	−2.25
Total costs	−9 142 841	−26.80
Net income	2 308 283	6.77
Net income	2 308 283	
Tax	−534 183	
Working capital	−257 033	
	1 517 067	
Net cash flow	1 517 067	
DCFR	10.8%	

Figure 8.8 An investment appraisal model

expectations are converted into a 'net present value' or NPV. The concept of NPV is used to balance the fact that an income of $1000 in one year's time is worth less than an income of $1000 today. There are two essential reasons for this. One is that $1000 today can be reinvested to generate additional income over the next year. The second is that $1000 today is a known quantity; the equivalent sum in a year's time will be worth less because there is an element of uncertainty about it.

Future sales revenues are converted into a net present value through the application of a discount rate. Thus, if a discount rate of 10% is applied, then an income of $1000 in one year's time has an NPV of $900. Similarly, an income of $1000 in two years time has an NPV of $810 and so on. In our example, NPVs have been calculated at a discount rate of 12%, and this yields total earnings from our investment project of about £11.5 million. For the same set of sales projections, different earnings will accrue, according to the discount rate adopted – the higher the discount rate, the lower the net present value of future sales.

The actual investment costs of the project are shown in the following lines of the summary report:

- Retail overheads show the ongoing additional costs associated with the improvement to the facility. These might include the extra staff costs of maintaining a larger store.
- Supply and distribution costs are higher, because more products need to be delivered in order to satisfy the extra demand for goods.
- Property costs reflect costs associated with the refurbishment, such as costs associated with extending or refitting the premises, or extra rent associated with the expansion.
- Third-party fees most likely relate to professional advice from architects, lawyers, accountants, planners or surveyors in association with the project.
- Credit card costs will also rise in proportion to the extra sales revenue.
- Taxation is naturally higher since greater profits can be generated.
- Working capital is a cost associated with short-term borrowings needed to finance the project.

Once all of these deductions are allowed for, the net cash flow from the project is still positive. The discounted cash flow rate (DCFR) provides a means of appraising the value of this cash flow relative to the money invested. If the NPV of an investment is $1000 and the net cash flow from the project (at current values) is $1200, then the DCFR would be 20%. The objective of the investment is usually to obtain a DCFR that exceeds some target internal rate of return (IRR) on capital, which would logically be closely related to the discount rate. In our example, we can see that the DCFR has been calculated at 10.8%, which is somewhat less than the 12% discount rate. There is a good chance that this project would not be deemed worthy of capital investment.

A more detailed discussion of modelling returns from capital investment is provided by Birkin *et al.* (2002). Many good management textbooks, such as Pike and Neale (1993), provide more detail on the financial concepts such as IRR, NPV, DCFR and the discount rate.

8.5 Providing a Decision-Support Capability

Once the models have been established, they are typically embedded within a Spatial Decision Support System (SDSS) that provides easy access to the model data, to various performance indicators relating to the modelling, and to the functionality of the model, within a software 'wrapper' that is easy to use and requires no depth of knowledge of either the software or modelling environments. Information is exchanged between the user and the SDSS, using a series of drop down menus and boxes that are readily familiar to all users of Microsoft Office products such as Word and Excel, and to all users of Windows-look-a-like products. A detailed discussion of the features and benefits of SDSS lies beyond the scope of the present work. For a wide-ranging review, the interested reader is directed to Geertman and Stillwell (2002), and to Birkin, Boden and Williams (2002) for a more specific discussion relating to the technologies described in this chapter.

The application of the SDSS for evaluation of retail business planning scenarios is demonstrated in Figures 8.9 to 8.11 and Plate 2. The first illustration shows that a particular study area has been selected, and a new store location identified. In simulating the opening of a new outlet, detailed information would usually be expected about the size of the outlet; its brand, layout, fascia (in effect, the store-type – many retailers adopt different store-types for different kinds of towns or market environments, see also discussion regarding format optimization in Chapter 9), parking, accessibility and any number of other store attributes that may influence the attractiveness of the outlet as a destination for customers. In addition to opening a new outlet, it is equally straightforward to simulate the impact of a store closure, upgrade, relocation or re-branding.

A key output from the modelling process is, of course, the sales forecast for our new store. This is illustrated for the new Middlesborough store in Figure 8.10. We can see that this new store location appears to be highly promising, and is expected to generate sales in excess of all but two of the existing supermarkets in this region, from an outlet that

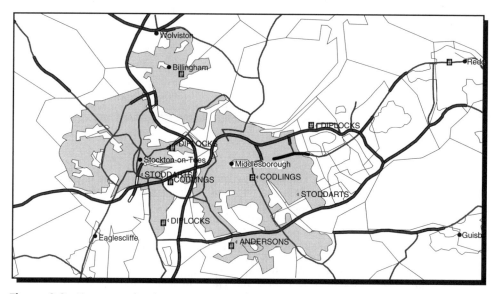

Figure 8.9 New store development scenario for Middlesborough: click on location to select

Scenario Name:	Middlesborough					
Description:	New store development					
Date:	16/04/99					

Store	Scenario	Size	Baseline sales	Scenario sales	Change	%Change
TAYLORS - Thornaby		38 441	531.7	485.4	−46.4	−8.7
KELLEYS - Middlesborough		23 600	184.9	141.9	−43.0	−23.3
DIPLOCKS - Thornaby		44 557	367.4	328.9	−38.5	−10.5
DIPLOCKS - Middlesborough		37 875	388.6	353.8	−34.8	−9.0
KELLEYS - Ingleby Barwick		28 200	266.4	233.5	−32.9	−12.4
STACEYS - Middlesborough		33 143	443.6	422.8	−20.8	−4.7
KELLEYS - Guisborough		17 200	187.8	170.1	−17.7	−9.4
STODDARTS - Middlesborough		16 600	63.1	50.7	−12.4	−19.7
ANDERSONS - Hemlington		6 750	58.2	46.0	−12.1	−20.9
TAYLORS - Middlesborough		45 000	288.4	276.3	−12.1	−4.2
DIPLOCKS - Stockton On Tees		47 579	496.5	485.2	−11.3	−2.3
CODLINGS - Middlesborough		10 000	100.6	89.4	−11.2	−11.1
KELLEYS - Stokesley		11 000	75.8	64.7	−11.1	−14.7
New Middlesborough Store	Open	27 200	0.0	525.0	525.0	100.0

Scenario Name:	Middlesborough				
Description:	New store development				
Date:	16/04/99				

Store	Scenario	Size	Baseline sales	Scenario sales
New Middlesborough store	Open	27 200	0.0	525.0
Fresh foods		6000	0.0	98.5
Packaged groceries		13000	0.0	245.9
Bakery		2000	0.0	38.2
Frozen		3200	0.0	75.2
Health & beauty		2000	0.0	25.1
Liquor & tobacco		1000	0.0	42.1

Figure 8.10 New store sales forecast

is much smaller than many of its competitors. Equally satisfactory is the impact analysis of Figure 8.11, which shows that less than 10% of the revenues are deflected from an existing store in the Middlesborough area, whereas the remainder of the new business is captured from competing stores. An important point to note from these illustrations is that it is the net sales of £384 200 per week (from Figure 8.10) that needs to be used as the basis for any cost-benefit calculation for this store, and not its own sales of £425 000 (as in Figure 8.11). It is these kinds of network impact that traditional models, such as the GIS and catchment-based models (see Chapter 6), typically find impossible to capture.

Further geographical detail is provided in Plate 2, in which we can see the market penetration provided by the new outlet. Here, 'market penetration' refers to the proportion

Scenario Name:	Middlesborough				
Description:	New store development				
Date:	16/04/99				

Brand	Size	Baseline sales	Scenario sales	Change	%Change
ANDERSONS	47 210	243.6	226.9	−16.7	−6.9
CODLINGS	30 000	285.8	265.9	−19.9	−7.0
DIPLOCKS	130 011	1252.5	1168.0	−84.6	−6.8
DULEYS	72 831	518.3	472.4	−45.9	−8.9
GILKES	24 000	219.4	211.1	−8.3	−3.8
KELLEYS	178 800	1538.8	1390.9	−147.8	−9.6
STACEYS	60 343	443.6	947.8	504.2	113.7
STODDARTS	30 600	113.5	98.8	−14.7	−13.0
TAYLORS	83 441	820.1	761.7	−58.4	−7.1
WHITTAMS	39 208	113.6	105.3	−8.3	−7.3

Figure 8.11 New store impact analysis

of customer expenditure within a particular small area (postal sectors in this case) that are captured by a retailer. More than one retail outlet may contribute to the penetration achieved within a single area. The map shows the penetration of 'Staceys', arising from a combination of two stores – an existing one, located in Thornaby to the east of the region, and the new store located in Middlesborough itself. Among other things, this map may provide an indication of the areas that might be targeted with local advertising and promotions following from any subsequent opening of the new store.

As we have seen in the preceding text, the model operates by simulating how customer patterns change for the network reconfigurations. The impacts on sales and market share are evaluated and provide valuable information to the investment appraisal process. The business model allows the impacts of various scenarios to be viewed: by outlet, by geographical area, by brand and by-product line. As we have argued above (see Chapter 7), the capability of customized decision-support models described in this section exceeds the capability of both GIS and old-fashioned 'gravity' models by some distance. The bottom line is that increased flexibility allows the models to better reproduce observed behaviour, and to provide better 'whatif?' predictions. A 'standard' model assumes that the attractiveness of a cluster of financial services branches is equal to the sum of its parts. The addition of a new branch increases the attractiveness of the cluster and therefore tends to suck a lot of new revenue into the cluster. The capture of existing business from other branches in the cluster is split evenly between competitors. On the other hand, the enhanced model assumes very little impact on the flow of products at the centre level. Most of the new business is generated via competition within the cluster, and of this, a disproportionate amount comes from an existing facility. More important than the numbers in this illustration is the fact that the parameters within the model may be adjusted to provide a balance between centre size, branch attractiveness and brand loyalty, which is appropriate to any particular market environment, and which can be calibrated to that market environment using known data.

In the next section, we show how the modelling framework can be applied to an entire reorganization of retail outlets in a particular city. The example relates to a major motor manufacturer and distributor in Austria (although the example is typical of a pan European reorganization). It was also designed to support senior managers keen to bring radical reform to their European operations.

8.6 The Development of a Cross Channel Management Plan for the City of Vienna for a Major Automotive Manufacturer

As we have described in Chapter 3, the automotive industry is characterized by a retail distribution system that has largely remained unchanged for the last 75 years. Virtually all manufacturers developed their distribution system by appointing franchised dealers, who were awarded a dealer area of responsibility in return for exclusive distribution rights for the brand they were representing. From the manufacturers' perspective, it made sense on a historical basis to appoint as many distributors as possible to maximize their presence in any particular market. Whereas this model worked successfully for a number of decades, in recent years, there is plenty of evidence to suggest it is no longer appropriate. Margins on new car sales across Europe average at about 1.5%. Return on capital employed (ROCE) across major volume manufacturers fluctuates between 6 and 8%, well below that of comparable retail sectors and wholly unacceptable to city investors. In many metropolitan markets, the plethora of dealers representing the same brand generates inter-dealer inter-brand competition. Customers in this environment have been able to visit a number of dealers in relatively close proximity and negotiate the best deal by trading one dealer's offer against another.

The economic problems of the existing dealer system found in most European countries have not escaped the attention of major manufacturers. In this section, we report on a major manufacturer who decided to address the issue of over-representation and duplication head on. The starting point was the recognition that they, as a company, provided a variety of different products through different distribution channels to a variety of different customers (see Figure 8.12). Translating this to the automotive business context requires the recognition that a traditional dealership provides a wide range of different products and services – for example, information about new and used products, the selling of such products, and servicing of both young and older used cars. At the same time, different types of customers have different types of requirements in using these products and services. Figure 8.12 identifies the different types of distribution channels that can be used to match the products and service requirements of different types of customers. These range from traditional full facility dealers, service workshop satellites, sales only outlets and the Internet.

To enable the format to be developed that would allow significant cost savings, reduplication and improvement of customer service, the customer market approach (CMA) has been used. Instead of a typical metropolitan market being divided into several franchise territories and having the various appointed dealers competing with each other for customers' business, a single CMA, in which the whole of the physical and virtual channels are owned by a single representative, effectively eliminates any intra-dealer competition. Consumers are effectively restricted in their choice of the retailer supplying new or used car products, but there are also significant potential consumer benefits. Historically, customers have travelled relatively long distances to purchase a new car, but subsequent

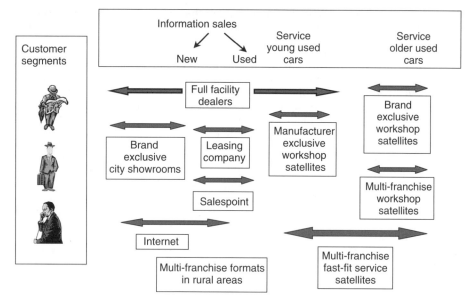

Figure 8.12 European automotive distribution channels

servicing and repair of the vehicle are usually undertaken at a more convenient and local-ized outlet. In the CMA approach, a customer can buy a new or used vehicle at his or her point of convenience, and also have it serviced or repaired at a similarly convenient location if under the 'ownership' of the same dealer. From the CMA operator's perspective, there are significant economies of sale to be generated from this approach. For example, the number of sales points within the metropolitan market can be reduced and the number of duplicated facilities, such as body shop, parts storage, vehicle storage and preparation centres can be substantially slimmed down. The CMA approach also allows a single con-sistent proposition to be presented to customers. Pricing, for example, of both new cars and servicing will be consistent across all sales and service points throughout the CMA. Furthermore, a customer can purchase a new or used vehicle at any sales point within the CMA, and also have it serviced at any service point in the CMA and still be 'owned' by the same dealer. The CMA concept also allows for consistent local marketing – instead of deal-ers competing with each other in the local newspapers and local radio, a city-wide offer is made from a single dealer. Another potential advantage of a single CMA approach is that it enables new region-wide initiatives such as Internet, car rental and retail adjacencies to be developed. With dealers competing in a single market, it is difficult for these opportunities to be developed.

There are potential problems and risks associated with these new large CMAs. The first potential issue is that if introduced they allocate a significant amount of sales volume to a single owner. If this owner underperforms, then the manufacturer risks a significant decline in market share. The preferred way forward, therefore, is for the manufacturer to take a minority stake in the operating company that owns the CMA and, through this equity holding, ensure that performance targets are met with potential cancellation of the ownership agreement if they are not. Another risk lies in the fact that most dealers have little experience of operating multisite facilities. The expanded CMA concept will require new management structures and operating controls to ensure effective implementation.

We now describe how the CMA approach, coupled with spatial interaction modelling, has been developed in the city of Vienna. Figure 8.13 shows a map of the existing main dealers and sub-dealers in Vienna. Vienna is currently divided into four dealer territories (with territory 4 having two main dealers) and a total of five main dealers and 17 sub-dealers within these territories. The manufacturer's current market share by small area in the Vienna region is shown in Plate 3 and demonstrates considerable variation, approximately between 5 and 15%. Table 8.2 shows the percentage of sales that each dealer makes within his own territory and within the other territories in Vienna. As can be seen, in most cases, only about half of the total sales that any dealer achieves are from customers who live within the dealer territory. This suggests that the current territory dealer boundaries do not act as natural barriers to customer flows – effectively, there is a single market for vehicle purchase within the city and this lends significant support to the idea of a single CMA.

Having established that the Vienna region operates as a single market and, thereby, demonstrating that the division of the region into a series of smaller territories is coun-

- Vienna city: Existing network
  - 5 Mains, 17 Subs

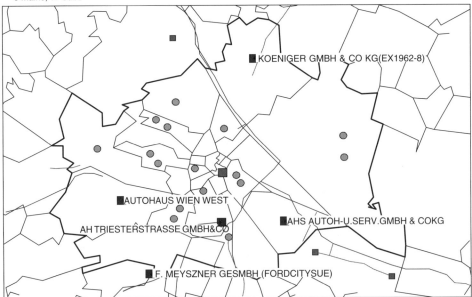

Figure 8.13 Representation in the Viennese automotive market

Table 8.2 Dealer containment in Vienna

CMA structure

- CMA sales flow analysis

Dealer	CMA location	Percentage of retail sales made in each CMA				
		1	2	3	4	Other
Wien West	1	50.2	6.7	5.0	34.1	4
Koeniger	2	20.6	58.6	5.2	13.2	2.4
AHS AutoH	3	17.4	20.2	37.5	21.7	3.2
Triessetrasse	4	23.1	9.2	8.2	52.9	6.6
City Sud	4	6.7	4.3	3.7	83.4	1.9

- This suggests that CMA boundaries do not act as natural barriers to customer flows

- Supports the single CMA approach

terproductive, our task was to review the representation and make recommendations on how a representation plan for the future within a single CMA environment could be developed.

The current sales network has 12 main dealers and two branches covering the Vienna super CMA (of these there are the five central Vienna dealers described earlier). For the new car sales representation, we examined two scenarios using the models. The first involved the closure of 3 of the 12 main dealers and the closure of the two branch dealers. In addition, we relocated one of the central Vienna dealers from the existing location to a site at Shopping City Sud, which is a major retail mall, approximately two kilometres from the existing dealer location. The second scenario is similar to the first, but with the additional closure of a central Vienna dealer. Table 8.3 summarizes the model results for the scenarios tested (with scenario 1 effectively being the current situation), and we can see that with the simple closure of the outlets the sales fall by 851 units (scenario 2) and 650 units (scenario 3). However, it is assumed that through the single CMA concept, it would be possible to improve the attractiveness of the dealer locations, both in terms of the brand overall and through additional investment in outlets and in marketing. It can be seen that scenario 2, with only 9 of the 12 major dealers remaining in total, achieves the CMA retail planning volume (7500), with only a 25% increase in outlet attractiveness. With a further reduction to 8 outlets (scenario 3), a 75% uplift in outlet attractiveness is required to exceed the sales target. Senior managers have recommended, therefore, that scenario 2 should be adopted as the new car sales representation plan.

We then proceeded to analyze the service business element of the client's activity in the CMA. Currently, both the 12 main dealers and the two branches provide service facilities along with 36 other service-only outlets. The business model was used to assess 3 scenarios. The first included keeping the 9 sales outlets proposed from the sales modelling above, but the retention of only 27 out of the 36 service-only outlets. The three

Table 8.3 Vienna scenario results

15 minutes

Current network - scenario 1			Scenario 2			Scenario 3		
Outlets	Ave drive time	% Coverage	Outlets	Ave drive time	% Coverage	Outlets	Ave drive time	% Coverage
25	2.9	98.5	9	4.6	93.4	8	4.9	93.4

20 minutes

Current network - scenario 1			Scenario 2			Scenario 3		
Outlets	Ave drive time	% Coverage	Outlets	Ave drive time	% Coverage	Outlets	Ave drive time	% Coverage
25	2.9	99.6	9	4.6	98.5	8	4.9	98.5

25 minutes

Current network - scenario 1			Scenario 2			Scenario 3		
Outlets	Ave drive time	% Coverage	Outlets	Ave drive time	% Coverage	Outlets	Ave drive time	% Coverage
25	2.9	99.8	9	4.6	99.1	8	4.9	99.1

30 minutes

Current network - scenario 1			Scenario 2			Scenario 3		
Outlets	Ave drive time	% Coverage	Outlets	Ave drive time	% Coverage	Outlets	Ave drive time	% Coverage
25	2.9	100.0	9	4.6	99.8	8	4.9	99.8

Scenario 1   **Current network of 25 outlets, 5 mains and sub dealers**
Scenario 2   **Takes the proposed 6 sales points with the addition of Wien 12, Wien 16 and Wien 22**
Scenario 3   **Removal of the additional location at Wien 12**

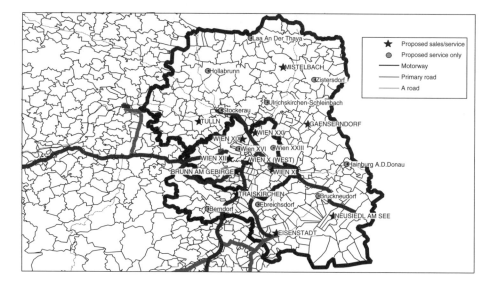

Figure 8.14   Representation in Vienna – scenario

main (and two branches) closed for sales facilities also become service-only points. In the next scenario, a further 16 sub-facilities are closed because of low potential. In the final scenario, a further 7 points are removed, owing to overlapping catchments. After a substantial amount of testing, the final scenario was chosen as the best plan for the client's network. The proposed network provides 91% of the population with a service point within a 20-minute drive time. This compares well with the existing 50 outlets, which give 95% coverage within the same drive time. The configuration of sales/service and service only outlets is shown in Figure 8.14.

The projects demonstrated that a considerable reduction in physical representation and distribution (and therefore costs) could be removed from the Vienna markets, without a substantial impact on customer accessibility to the client's outlets. These proposals are now being implemented in Vienna.

8.7   Conclusion

It has been argued that developments in spatial interaction modelling have been significant over the last twenty years. Unfortunately, much of this goes unrecorded in the retail trade press (although it is, perhaps, reassuring to see something of a resurgence in interest at the new 'The Art of Store Location' conferences held annually in London by Henry Stewart Conference Studies, see e.g. Bond 1997). There is a great deal more experience in using the models and understanding the limitations and problems associated with them. The models have been applied to a wide range of business scenarios and we have demonstrated a range of applications in this chapter. It is unlikely these days that models will simply be used to investigate the impacts of new store openings. A number of retailers are using this technology to reorganize entire networks as new forms of distribution become more important. It is also true to say that these models are increasingly used alongside GIS and geodemographic systems (Chapters 6 and 10), optimization models (Chapter 9) and, let us not forget, gut feeling and intuition. Gut feeling and modelling do not have to lie at either end of the spectrum: the two can go hand in hand to provide real decision support systems for retail clients.

NETWORK REPRESENTATION PLANNING

9.1 Introduction

Some years ago, it was determined that the customary practical examination for driver licensing ('the driving test') within the United Kingdom would be supplemented by a written examination. Each candidate would be obliged to pass the written examination before moving on to the practical test. In order to support this change, the Driver and Vehicle Licensing Authority issued a Call For Tender, asking for recommendations about the location of new Examination Centres for the written component. Every candidate should have access to a centre, subject to the following rules:

● Access within 15 minutes driving time in an urban area (population density of 1000 plus per square kilometre)
● Access within 30 minutes driving time in a suburban area (population density of 500–1000 per square kilometre)
● Access within 45 minutes driving time in a rural area (population density of less than 500 per square kilometre)

The question naturally arises: how many centres are needed, and where should they be located?

The Examination Centre problem may be solved using an iterative relocation algorithm with zone penalties (which can be applied when the threshold rules are violated), which is very similar to the territory planning models introduced in Chapter 6. The precise solution to this problem depends on the level of detail adopted for the analysis of population density and accessibility (unique addresses, unit postcodes, electoral wards...; cf. Section 6.2) and of the way in which drive-times between these locations are estimated. A potential solution is shown in Plate 4. In this example, the population has been segmented into approximately 9000 postal sectors. Each of the sectors has been categorized according to the population density rules specified in the preceding text. For each postal sector, we have identified the drive times to every other postal sector within 60 minutes. Driving test centres are then located in postal sectors, so that we arrive at the smallest number of test centres from which every postal sector can be covered within the specified rules (i.e. each sector classed as rural has a test centre within 45 minutes and so on). The solution shown in Plate 4 achieves this with only 120 test centres. It can be seen that most of the sectors are of a regular shape, as one would probably expect, although some can tend to be elongated through the effect of major roads along which the drive-times will tend to be shortened. Because of the population density rules, the centres tend to be smaller in urban areas such as London, Birmingham and Manchester.

Figure 9.1 illustrates the location types and the distribution of capacity between them. Figure 9.1(A) shows the distribution of potential customers (i.e., the total population) across the centres. It can be seen that there is enormous variation, from centres that serve areas with less than 50 000 people, to those with more than a million. Further detail is provided in Figure 9.1(B). As one would expect, all of the centres with less than 100 000 potential customers are located in rural areas. On the other hand, the majority of centres with over a million customers are urban in type.

Suppose a further constraint is introduced that says that no centre may serve a population in excess of, say, 750 000 people. This constraint is violated by twelve of the existing centres. Without describing further solutions in detail, we can see that this can be

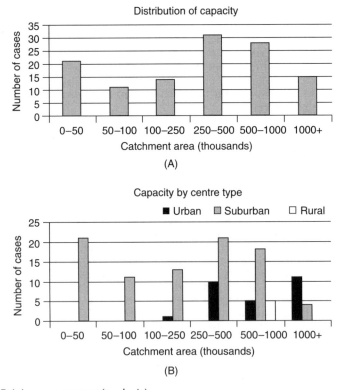

Figure 9.1 Driving test centres (analysis)

achieved with at most 132 centres, as long as none of the capacities exceeds 1.5 million (by splitting each of the violators into two centres rather than one).

The capacitated Examination Centre problem may be seen as a relatively complex network-planning version of the territory assignment problems introduced in Chapter 6. It is relatively complex because there are many centres, and because there are multiple rules about density and accessibility. Nevertheless, it is also simple in one important respect, which is that all potential 'customers' behave deterministically according to well-specified rules. To put it another way, it is the supplier (the DVLA) that provides the rules about where individuals must go to do the examinations. While this is realistic for certain types of services – generally those that are provided by public authorities – an extra dimension is added if consumers are allowed to exercise choice of location. Suppose, for example, a British supermarket wishes to enter the grocery market in France, and to open 100 new stores (cf. Chapter 4, Section 4.3); from which locations can it generate the biggest customer base?

The problem of optimal location with consumer choice, as faced by our imaginary supermarket, has an extra dimension of difficulty. In order to evaluate any potential solution, the behaviour of individual customers must be predicted using something like a spatial interaction model (see Chapters 7 and 8). This problem appears to have been recognized first by Hodgson (1978). Problems of realistic size have been addressed using a procedure known as the 'Idealized Representation Plan' (IRP), which has been documented by Birkin, Clarke and George (1995). Further discussion of the IRP is provided in Section 9.2.

It is difficult to overemphasize that the IRP is an extremely hard problem in both computational and theoretical terms. As we saw in Chapter 6, the Travelling Salesman Problem (TSP) is a famously difficult computational problem when the salesperson needs to visit more than about 20 places in the correct order. The IRP has many similar features, but, typically, involves not only the choice between many thousands of locations, but also the interactions between individual pairs of thousands of locations. Birkin *et al.* (1995) showed that to solve a typical problem required approximately three months processing time on a high specification Sun workstation. Although George *et al.* (1997) found that this time could be dramatically reduced using parallel computer architectures, neither the software nor the hardware required to achieve this is easily available. However, the effort necessary to overcome these issues is well spent, because of the importance of this class of business planning problems. The following sections are designed to give examples of both the range of business planning issues and the ways in which they can be addressed using network-planning technologies.

9.2 Case Study 1: Automotive Network Optimization

9.2.1 Background

At first sight, the car industry does not appear to be the most natural application area for location planning techniques. Surely, motor vehicle sales are determined by testosterone, branding, lifestyle, safety and panache! In fact, relative location is as important as any of these factors, and more important than most. Consider the case of Renault, for example, which in the United Kingdom is a middle ranking, mid-market marque. In the top one hundred towns where a Renault dealer is located, the manufacturer enjoys a 10% market share. In the top hundred towns lacking a Renault franchise, the share falls below 5%.

Furthermore, the location decisions faced by car manufacturers are particularly acute, for a number of reasons. Most car dealers in Europe and the United States are franchised. Therefore, a manufacturer has to try and build market share without direct control of its own sales channels. In many countries, notably the United Kingdom and United States, the hegemony of the manufacturers is seriously threatened by the growth of strong independent dealer groups. Many other national markets are over-dealered, with too many franchises operating at low or negative profitability. Combined with the consolidation of manufacturing into a small number of mega-international brands (i.e., Ford-Volvo-Jaguar; Daimler-Chrysler-Benz) there is both the opportunity and necessity for major network restructuring. The ability of dealers to control exclusive geographical territories is increasingly under threat and the promotion of multiple marques from a single dealer location ('multi-franchising') is increasingly common.

Thus, the automotive sector provides a fertile arena for market planning activities. Major international companies, such as RL Polk and USAI, provide automotive market analysis globally, whereas companies such as GMAP, and CACI have a strong European and UK presence. In Chapter 6, we have already looked at some applications from the perspective of territory planning; in subsequent sections of this chapter we will consider network planning applications.

9.2.2 Idealized representation planning

Suppose a new car dealership is located on a greenfield site. The likely sales of the new dealership may be estimated by considering the number of customers in its environs, their buying preferences and behaviour, and the distribution of competing automotive franchises that would seek to penetrate the same market. In other words, a spatial interaction model will provide a highly appropriate framework for estimating dealer performance.

Now suppose a new manufacturer wishes to create a whole network of dealerships, say 100 in the United Kingdom. Not only is it important to select 100 good locations but also each of these is 'interdependent' from the other locations. Therefore, even though there may be fifty good locations in London, only five or six of these may be required, whereas in Leeds only two of ten or twenty good locations may be needed. The locations must be assessed, not only individually but also in relation to the coverage and competition between each another. This is the IRP, which we began to introduce in Section 9.1. We can write it down in non-mathematical terms:

Find a network of N outlets so that this network sells more cars than any other network of N outlets, assuming that sales within the network may be predicted using an appropriate spatial interaction model.

It is also possible to generate versions of the IRP in which the number of outlets to be found is also a component of the problem: for example, to find the best network of outlets to deliver a certain market share or sales target. However, these alternative formulations are, in essence, a special case of the specification provided above. Thus, to find a network that delivers 10% market share, one would probably begin by estimating the number of dealers required, then running the IRP. According to results, the estimate of the number of dealers required would be updated and the IRP run again: and so on, until a satisfactory solution is reached. Note also that it is possible to specify the IRP so that an optimum level of profitability is achieved. In principle, this is straightforward and looks attractive, but, in practice, this requires that costs be estimated at each location, which can be difficult. Furthermore, in the car market dealerships are typically franchised, as we have already seen. In which case, assuming that car prices are relatively fixed, as they generally are, then manufacturers really do want to maximize their sales and the cost of doing business is a concern for the franchisee.

An objection that is sometimes raised against Idealized Representation Planning is that, usually, it is impossible to adopt a carte blanche approach to planning. For example, if an established retailer wants to find the best locations for his/her 200 stores, then surely she/he is somewhat constrained by the investment that has already been made in the network. This argument is less strong in the automotive sector for a number of reasons. Firstly, as we have just seen, distribution is franchised, with the manufacturers generally having most power. Therefore, manufacturers are relatively free to massage their networks if they feel it is appropriate. Secondly, many manufacturers are genuinely moving into new geographical markets. In particular, established brands such as Ford have begun to open up new markets in Eastern Europe and, to a lesser extent, Africa and the Middle East in recent years. Completely new brands, such as Kia and Proton, have also come into play and clearly demand new networks, whereas the introduction of new products and branding strategies, such as the compact Jaguar and SMART vehicles, provides an impetus for innovative thinking. Finally, the changing nature of both demand and supply in the automotive sector has led to significant changes in market planning, for example, larger dealer areas of responsibility and product specialization (Financial Times, 1998).

Following our description above, the IRP may be viewed as an embedded spatial interaction model. In essence, the trick is to run a spatial interaction model millions of times with different dealer locations, in such a way that the best solution can be found. The IRP, therefore, has the same data requirements as a conventional spatial interaction model. In the case of the car market, availability of appropriate data is typically very good. In many countries, vehicle licensing agencies, such as Society of Motor Manufacturers and Traders (SMMT) (United Kingdom) and AAA (France) and others, collect vehicle registrations at small area level by marque and model. Thus, in any of these countries, one can find the number of Ford Mondeos, against VW Golf, Audi A4 or any of a variety of competing models, at postal sector or equivalent small area. When combined with dealer lists, which can typically be obtained from individual manufacturers, an excellent platform for modelling is provided.

The structure of the IRP heuristic is described by Birkin, Clarke and George (1995). Imagine that an ideal network of 100 dealers is to be found. Initially, one would find the single best location for a dealer. This is done by imagining a car dealer at every possible location (e.g. every postal sector) and running the associated spatial interaction model to calculate dealer sales. The location that generates the highest sales is selected as the first location. This dealer now becomes fixed, and the procedure is repeated to find the next best location. The first dealer is then 'perturbed' (moved around) to check that it is still in the best place. This procedure is repeated until 100 locations have been identified.

Two example IRPs for the country of Denmark are shown in Figure 9.2. The objective has been to take an existing network of dealers and to optimize the network, so that the same number of cars can be sold through fewer outlets by finding the ideal locations. This process is illustrated for two manufacturers, one, a major manufacturer with more than 100 dealers in the existing network (see Fig 9.2a,b) and the other a middle ranking manufacturer with around 80 dealers in the existing network (see Fig 9.2c,d).

A powerful feature of the IRP is that it is able to incorporate existing network constraints – for example, if we want to create a new network of 100 dealers from a combination of 50 existing dealers and 50 new ones. This is an important consideration for practical applications, as we shall see in the next section.

Three problems with the IRP heuristic described above are the quality of the solutions, the speed of the algorithm, and the practical deployment of the outputs.

9.2.3 Solution quality

The IRP heuristic is unlikely to produce poor solutions to the network optimization problem. Equally, however, it is well known that hill climbing procedures such as this will rarely yield perfect solutions to complex problems. This has been demonstrated for sample IRPs by Birkin *et al.* (1995). In the case of the TSP, which was discussed in Chapter 6, better solutions are found using a 'simulated annealing' algorithm (Press *et al.*, 1989), but, again, this technology is insufficiently powerful for the IRP. George *et al.* (1997) have described an alternative solution procedure, that uses a genetic algorithm (GA) to find improved solutions. The GA does not work with a single solution, as with the conventional IRP heuristic; rather, it works with a population of different solutions that are refined and combined in such a way that progressively better solutions are allowed to 'evolve'. For more details, see Holland (1975), Davis (1991) and George *et al.* (1997).

9.2.4 Solution speed

To implement the IRP for a reasonable number of dealers against potential locations requires the evaluation of a complex spatial interaction model hundreds of thousands or millions of times. It was shown by Birkin *et al.* (1995) that to do this for the United Kingdom in order to allocate several hundred dealers between several thousand postal sectors would require approximately three months processing on a Sun Sparc 20 workstation. Even with continued advances in computation, this workload remains burdensome in the extreme.

George *et al.* (1997) demonstrated that speed increases by a factor of 2700 times could be achieved through the application of massively parallel computing power. Because GAs work with a population of solutions, these require even greater computing power, and hence, are only facilitated through the application of parallel computing technology. Unfortunately, neither parallel computer hardware nor the software required to exploit it, are generally available to network planners. Certain features that are specific to the IRP in both its heuristic and GA version have recently been exploited to allow solutions to be evaluated much more rapidly. Nevertheless, computational constraints remain a considerable bottleneck in the production of representation plans.

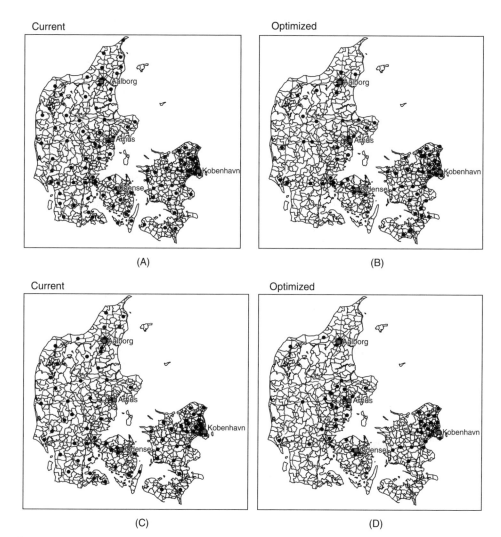

Figure 9.2   Denmark IRP

9.2.5   Practical deployment

If Idealized Representation Plans are to prove useful to planners, then they must clearly be capable of deployment. Whether the network to be managed is already established or completely new, it will be necessary to manage the plan as a living and dynamic entity, and not as a lifeless and static blueprint. For example, an 'ideal location' cannot be commandeered – good sites, which may or may not be close to the ideal, will need to be acquired when they become available.

For all of these reasons, the application of ideal network planning generally rests on the ability to break a network down into manageable chunks. It is to this purpose that the concept of Containment Areas has been developed, and to which we now turn.

9.2.6 Containment areas

One of the fundamental principles of network planning is 'interdependence', the effect of doing something at one location is dependent on what is happening at neighbouring locations. Consider the simple example shown in Figure 9.3. Floorspace is available at the White Rose Centre. Retailer A already has stores in Leeds, Bradford and Wakefield; Retailer B has no existing stores at these locations. A new store for Retailer A will have a significant impact on its existing trading patterns; Retailer B has no such problems. It would be unwise for either retailer to base the decision solely on the characteristics of the White Rose Centre, and without consideration of other stores in the locality.

On the other hand, it makes equally little sense when evaluating the White Rose Centre to consider Aberdeen or Exeter. These stores are far enough away to have no impact. If a fixed set of regions is created, the problem of irrelevance can be eliminated, but there are potential 'boundary problems' at the edge of each region. One way to handle this conflict is to define a set of regions that are largely homogeneous and self-contained, so that boundary effects are minimized.

To fix ideas, let us suppose that the set of journey-to-work flows between small areas is known. This information is routinely collected via the census of population in many countries. The problem is how to partition these small areas into regions so that the flows

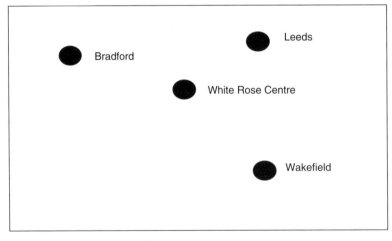

Figure 9.3 Retail locations in Yorkshire

within regions are maximized, and the flows between regions are minimized. This problem has some similarity to the derivation of 'Functional Regions' at the Centre for Urban and Regional Development Studies (CURDS) at Newcastle in the early 1980s (e.g. Coombes *et al.*, 1986).

The problem is qualitatively different to those of Sections 6.2 to 6.4, since it involves connectivities between small areas rather than just scalar attributes of those areas. However, if one thinks of the journey-to-work flows as proxy drive times between areas, then the problems start to look more similar. In Section 6.4, we wanted to find out the least possible number of salespersons to meet given service levels for a client base. In this section, the objective is to minimize the number of regions to meet a given containment level. A containment area solution for the Stockholm region is illustrated in Plate 5.

A strategic application of containment areas has been described in the Financial Times (1998). In this report, Ian McAllister, Chief Executive of Ford United Kingdom, explains how the concept of the Customer Marketing Area (CMA) has been applied across the whole of the UK dealer network. In part, the rationale for this has been to eliminate destructive 'Intrabrand competition' between dealers. Intrabrand competition arises from the proliferation of dealerships within relatively small geographical areas, and can lead to fragmentation of the brand (see Chapter 8 for Vienna example). To recap, in many areas customers may wish to purchase a car from a local dealer near to where they live, but to have the car serviced close to where they work, shop, or play golf. However, this could mean buying a car from one dealer, and having it serviced by another dealer. This kind of competition for business is natural for the dealers, but damaging to the Ford brand. Another important effect is that many European markets have become over-dealered, with very low or even negative profitability. Ford have therefore utilized the CMA concept as a means of identifying larger DARs in a structured way. This has allowed the Ford brand to be projected more consistently across a stronger network of dealers. Although the number of dealerships may have been reduced, this is not necessarily true of customer access points, since dealers have been encouraged to diversify their networks within a single CMA. Thus, after-sales support may now be provided through dedicated 'service-only' satellite points, at more convenient and distributed locations than the main dealer sales points (again see Chapter 8).

The Containment Area approach is distinctive with respect to other regionalization approaches in at least two respects:

1. It is a general-purpose approach that has been applied in a number of European countries. For example, Table 9.1 shows a pan-European regionalization, in which 19 countries have been split into self-contained regions on the basis of minimal cross-boundary interactions between the various regions.

2. It forms a basis for commercial applications, which are reviewed more extensively in Chapter 8, Section 8.5, and in Sections 9.3, 9.4 and 9.5.

In practice, Containment Areas are generated using a variety of data sources, which often includes retail interactions in addition to journey-to-work flows and data for specific businesses. For example, in defining a set of Containment Areas for a bank or building society, it would be appropriate to include information about account ownership, transaction behaviour, and ATM usage by the organization's own customers.

Table 9.1 European containment area solutions

Country	CMAs	Population	Area (Sq Km)	Population density (Pop/Sq Km)	Average population per CMA	Average JTW containment
Austria	24	7 795 786	83 641	93	324 824	77.6%
Belgium	29	9 941 896	30 513	326	342 824	74.5%
Denmark	20	5 213 472	42 805	122	260 674	80.3%
Finland	16	5 098 754	331 261	15	318 672	88.1%
France	124	56 738 473	547 135	104	457 568	82.4%
Germany	149	81 262 432	355 646	228	545 385	82.9%
Ireland	26	3 525 719	68 496	51	135 605	n/a
Italy	119	57 062 670	302 293	189	479 518	85.9%
Netherlands	43	15 344 668	34 872	440	356 853	80.1%
Norway	19	4 344 482	316 054	14	228 657	93.9%
Portugal	25	9 868 037	88 834	111	394 721	82.6%
Spain	83	39 433 942	505 575	78	475 108	82.1%
Sweden	26	8 816 381	438 937	20	339 092	86.8%
Switzerland	32	6 888 037	40 826	169	215 251	86.4%
UK	110	54 853 618	243 907	225	498 669	80.1%
Poland	49	38 127 514	308 711	124	778 113	n/a
Hungary	20	10 362 129	92 887	112	518 106	85.2%
Czech Rep	23	10 321 345	78 251	132	448 754	92.2%
Greece	18	10 313 687	131 505	78	572 983	

Note: CMA-Customer Marketing Area

9.3 Case Study 2: Financial Services Network Optimization

9.3.1 Background

In this section we switch from optimisation as a formal mathematical programming routine to optimisation as getting the network planning strategy correct.

This is probably the most dynamic of the service sectors currently, within the United Kingdom and elsewhere. On the one hand, the introduction of new channels such as telephone call centres, and more recently Wireless Application Protocol (WAP)/ Internet, have given providers an opportunity to deliver services at lower cost. At the same time, however, new entrants to the market (such as Egg and Goldfish) and the freer flow of information about competing products have forced down margins and profitability. Established banks and building societies have also found it difficult to convince customers and regulators about the benefits of reduced service provision at lower cost. Furthermore, the financial services sector is experiencing rapid consolidation, as pressure from shareholders and customers forces institutions towards mergers and alliances.

The distribution channel issues are considered at length in Chapter 10. From a network planning perspective, the following questions are all of interest:

● How can networks be planned to provide the right number and mixture of delivery channels?

● How can networks be integrated when two organizations are merged?
● How can networks be rationalized to provide effective delivery of services at lower cost?

Each of these questions will now be considered.

9.3.2   Representation planning for financial services

Once again, the containment area approach provides an excellent basis for network planning in this sector. This can be understood with reference to the local market example in Table 9.2. Column 1 shows the distribution of transaction volumes for residents of Wetherby in North Yorkshire. The volumes show a split between two destinations, the local financial centre of Wetherby, and the major regional centre of Leeds. Customers are attracted to Wetherby by its proximity; but also drawn to Leeds for employment, shopping, football, nightlife and other activities. It is important to emphasize that while discrete groups of customers may tend to utilize either centre exclusively, the majority will tend to patronize both Leeds and Wetherby on different occasions. There is a contrast between financial services activity, in which the relationship between customer and provider is maintained through multiple transactions over a long period of time, and retailing, in which each individual purchases represent a unique and complete transaction.

Table 9.2   Flows from Wetherby & Leeds

	Wetherby	**Leeds**
Leeds	4645	8196
Bradford	937	438
Wakefield	83	417
Pontefract	5	78
Castleford	46	41
Keighley	51	37
Ilkley	55	12
Wetherby	3680	94
Crossgates	55	33
Otley	18	86
Yeadon	46	41
Shipley	5	57
Guiseley	37	49
Garforth	23	70
Kippax	37	37
Pudsey	14	70
Headingley	37	49
Adel	9	78
Wyke	46	20
Morley	51	37
Horsforth	55	33
Roundhay	65	29

Table 9.2, Column 2 shows how things look from the Leeds end, where we see clusters from a whole series of 'Wetherby's. Therefore, we cannot determine what levels and types of provision are required in Leeds, without reference to the surrounding areas that provide so many of its customers. However, as in the discussion of Section 9.1, for example, one does not wish to plan provision in Leeds with direct reference to customers in London or Birmingham, or even Halifax or York. Thus, the containment areas can be used to provide a working definition of 'Leeds'.

The second step in planning representation for financial services is to generate a 'what if'? modelling capability. Figure 9.4 shows some simple examples for the case of Leeds, using a spatial interaction model with varying parameters. Low attractiveness and high proximity would most probably be associated with ATM transactions, which are frequent and would not usually be associated with long trips. High attractiveness and low proximity might be associated with mortgages, which are high-value, infrequent transactions, in which getting the right deal is likely to play a more important part than having the provider next door. Moderate attractiveness and moderate proximity might be associated with branch transactions or enquiries, which are regular, without being frequent, that is, once or twice per month.

The reader can visualize from the examples of Figures 9.4 that it is possible to produce interaction models that provide a good representation of customer behaviour for financial services. In practice, the models can be quite complex for a number of reasons. One is that, as we have seen in the preceding text, financial services relationships tend to be quite longstanding. Thus, many people may have relationships with providers or branches that were established many years ago and seem illogical with respect to existing

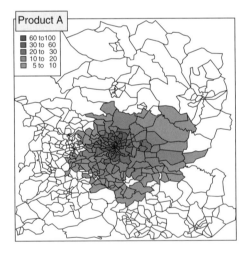
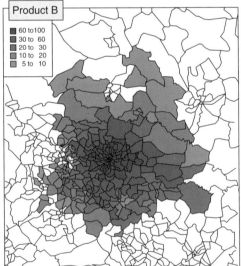

Figure 9.4 Model flows to Leeds

provision. A second reason is that financial services transactions, such as paying bills or withdrawing cash, are often 'distress' activities, in that they are unlikely to demand a specific trip, but will be done in association with something else, like a regular shopping trip. Therefore, the attractiveness of destinations may have more to do with other activities than banks. Thirdly, it is quite common to have multiple provision at a single location. For example, GMAP has identified just over 3000 'financial services centres' within the United Kingdom, which is a large number. At the same time, however, there are more than 20 000 ATMs, about eight per centre, many of which offer common services. The models must incorporate a high degree of subtlety in order to get the right balance of preferences. Further discussion about the modelling of financial services transactions is provided by Birkin and Clarke G (1998).

An important application of the models within representation planning is to produce comparative indicators or 'benchmarks' between locations. One of the simplest but most useful benchmarks is simply to calculate a 'catchment population' for each centre, which is the number of customers regularly using the centre. In our previous example, the catchment population of Leeds would be high and the catchment population of Wetherby would be low. Ranking the centres within a Containment Area, according to their size and the presence or absence of financial outlets, provides an immediate idea of candidates for the introduction of new outlets (large centres that lack provision) or rationalization (small centres with unnecessary provision).

Many other benchmarking indicators may be introduced to enrich the analysis process. For example:

Provision ratio is a measure of the service levels in an area, which takes account of spatial interaction. Thus in our earlier example, closing the Wetherby branch does not eliminate provision because customers can still travel to Leeds and other surrounding centres.

Average distance travelled is a useful measure, especially if disaggregated to focus on different customer groups. For example, Barclays has run into considerable PR difficulties at the latter end of 2000 through its plans to close bank branches in rural areas, often providing residents with no practical alternative branch.

Market penetration shows the relative importance of different providers across small geographical areas. Other things being equal, areas of low market penetration will provide the best opportunities for new development; areas of high penetration may provide opportunities for rationalization.

Measures of sales for each branch, and associated ratios, such as sales per head of catchment population and outlet performance (actual sales against model expectations), provide useful measures of the efficiency of each branch.

For further discussion of Performance Indicators, see Clarke G and Wilson (1994), or Birkin *et al.* (1994). Retail examples are discussed by Birkin and Foulger (1992).

A battery of indicators for a single containment area is shown in Table 9.3. On the basis of size alone, the centres of Pontefract, Castleford, Keighley and Ilkley appear to offer most new growth potential. However, Castleford and Keighley both have good supporting provision from neighbouring centres. Pontefract and Ilkley are probably much better choices, therefore. On the other hand, Kippax, Horsforth and Adel are all closure candidates. Horsforth has other providers, so residents would not be left entirely bereft should the MyBank branch withdraw. Adel has good provision from other local centres. On

Table 9.3 Financial centre performance indicators

	CatchPop	Outlets MyBrand	Comp	Share of sales	Outlets/ head	Av dist	Market pen	Prov ratio
Leeds	512 000	4	123	4	0.25	9	3	144
Bradford	286 000	3	76	5	0.27	8	4	159
Wakefield	109 000	2	24	10	0.24	7	8	140
Pontefract	50 000	0	12	0	0.23	6	1	113
Castleford	44 000	0	13	0	0.30	6	1	151
Keighley	36 000	0	11	0	0.31	4	4	157
Ilkley	35 000	0	6	0	0.16	7	6	75
Wetherby	32 000	1	10	12	0.34	6	10	179
Crossgates	28 000	1	2	34	0.11	3	22	82
Otley	27 000	1	6	16	0.25	4	12	125
Yeadon	25 000	1	3	25	0.17	3	15	119
Shipley	23 000	0	5	0	0.21	4	4	100
Guiseley	19 000	1	7	15	0.41	3	12	257
Garforth	18 000	0	9	0	0.47	4	1	294
Kippax	18 000	1	2	0	0.14	7	2	28
Pudsey	18 000	0	7	0	0.36	3	3	229
Headingley	13 000	0	7	0	0.56	2	2	343
Adel	11 000	1	1	0	0.18	2	2	173
Wyke	11 000	0	2	0	0.18	3	2	125
Morley	9 000	0	6	0	0.69	3	2	423
Horsforth	6 000	1	6	20	1.08	2	18	648
Roundhay	6 000	0	4	0	0.58	2	2	358
	1 336 000							

the other hand, residents of Kippax already have to travel long distances, so any closure here could be controversial and best avoided.

Performance indicators can be compared between regions and within regions. A useful technique is to combine pairs of indicators in the form of a Boston Chart. For example, a comparison of market share against network density is commonly used to focus development priorities. Figure 9.5 shows the distribution of market share against network density for a financial services organization across a number of marketing areas. Towards the bottom right hand side of the chart, high market share is achieved through a sparse network, and little action is likely to be needed. These locations are labelled as 'Type 1' areas. When high share comes from a large network, then there may be scope for some rationalization to increase sales per outlet, reduce costs, and increase profitability. These areas are shown as 'Type 2' at the top right of Figure 9.5. Where an extensive network generates low sales ('Type 3'–top left), then the network needs to be reconfigured to get the branches into better places, or to provide better facilities. When low share comes from a sparse network, then it may be that critical mass has yet to be achieved. Thus, in 'Type 4' areas, to the bottom left of Figure 9.5, the network needs to be extended to become more effective.

Once we have the ability to run models, and to benchmark existing provision, then model scenarios and forecasts can be constructed and evaluated. Further examples of this type are illustrated in Section 9.4 .

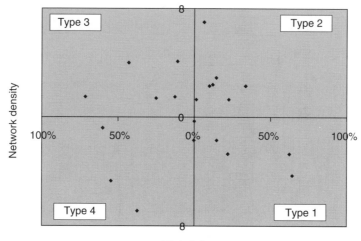

Market share

Figure 9.5 Boston Chart–Financial Services

Before leaving this section, it is also useful to note that the carte blanche representation plans discussed earlier are also of relevance to financial services. This is particularly true for new market entrants, including internet banks, which are expected to begin a process of physical distribution at the next stage within their development (i.e. so-called *bricks and clicks* or *clicks and mortar*).

9.4 Post-merger Network Optimization

Next, we focus on the use of the model outlined in Section 9.3 to investigate the idea of network integration. Here, we are concerned with identifying the optimum configuration of a new combined network of stores following a merger or acquisition. That is, given that the new combined network is very different from the two previously separate networks (in terms of both size and performance), and given the assumption (barring some billion to one chance) that the new network is not configured optimally, we are trying to find the optimum size of the network and their locations. That is, how many branches should we close, how many should we open/relocate, and which ones and where?

Before we perform a search for the optimum network configuration, first let us take a closer look at a real example. Here we use the new combined Barclays/Woolwich network following their merger in 1998. The new combined network contains 2292 branches and generates a market share of 14.5% in the combined UK retail current, savings, and mortgage account markets. For ease of analysis, we now focus on a subset of this network in the Kent CMA in South-East England. In this CMA, the combined network totals 67 branches and achieves a 17% market share of the three product markets, some 2.5% above

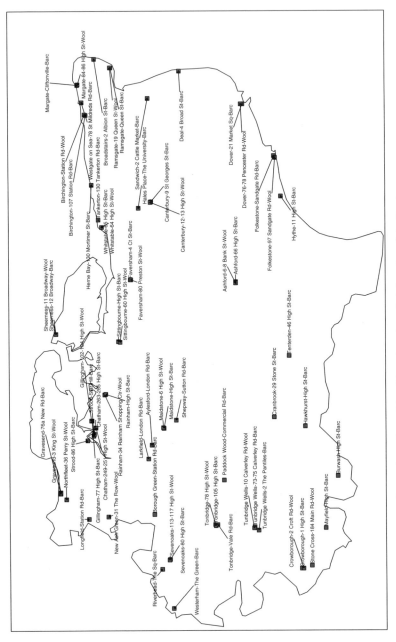

Figure 9.6 Existing configuration of combined barclays/woolwich network in Kent

its national average. The locations of each branch in the network and the distribution of market share is shown in Figures 9.6 and 9.7. We notice that in some localities, many centres in the new combined network contain more than one outlet. In fact, 24 outlets are located in centres that contain one or more outlets of the same brand, representing a 36% overlap in the CMA as a whole. In some cases, they are on the same street, and even next door! (See for example, the location of branches in Sheerness at the top of Figure 9.6). In terms of market share, we see that there are considerable variations throughout Kent. For example, in the West market share reaches 35% in some postal sectors, whereas in the South East it falls to less than 5%.

Barclays' planned strategy is to close all Woolwich outlets within 100 metres of an existing Barclays outlet. On the basis of these simple criteria, 10 outlets will be closed in Kent. Figure 9.8 shows the new distribution and performance as a result of these network changes, and Table 9.4 compares the Barclays strategy with a strategy of 'do nothing'. As we can see, the strategy will reduce costs (based on the obvious assumption that the cost of operating 57 outlets will be less than operating 67) and marginally improve network efficiency (by 0.61% – measured in terms of sales per outlet). Given these efficiency improvements, we can also assume a marginal improvement in network profitability. However, as we can also see, despite cost savings and efficiency improvements, the strategy has a negative effect on revenues and market share. Market share for the new combined organization decreases from 17 to 14.6% because of a

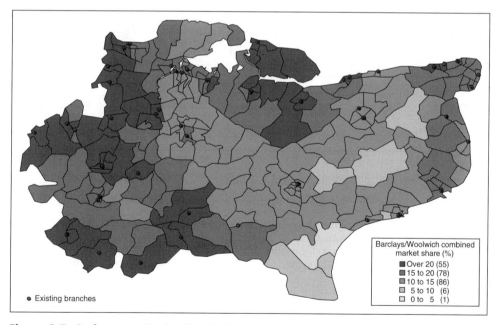

Figure 9.7 Performance (Market Share) of combined network in Kent

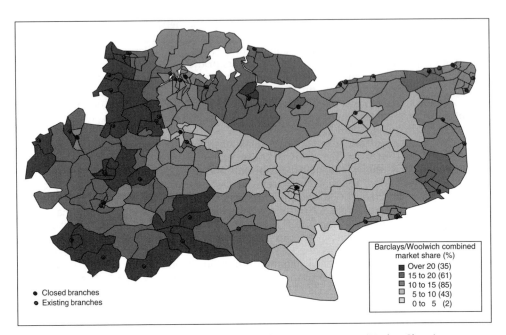

Figure 9.8 Performance of Barclays network configuration strategy (Market Share)

Table 9.4 Existing network performance vs Barclays strategy

Strategy	Outlets	Sales	Market share (%)	Sls/Outlet	% Change	
					Volume	Sales/Out
Retain existing	67	48 845.75	17.03	729.04	na	na
Barc strategy	57	41 809.39	14.58	733.50	−14.41	0.61

14% decrease in revenue volumes (versus a strategy of retaining the existing distribution). As outlined in Section 9.3, this decline in sales and market share mirrors evidence from other recent UK retail financial mergers. In fact, given the merger objectives of maximizing market penetration (measured in terms of revenues and market share), the strategy achieves the opposite of what Barclays intended. Again, the question is, therefore, what is the optimum configuration of the combined network? Could they have done better? Let us now, therefore, perform a search for the optimum combined network configuration.

As outlined in Section 9.3, there are a number of network configuration options that we can consider in a search for the 'optimum'. Here, we perform a number of searches, in which we consider just rationalization (closures), just opening/relocations, and closures and relocations/opening simultaneously. We begin by looking at alternative integration strategies in which we consider rationalizations only. Can we match the performance of the existing combined network with a reduced number of outlets and/or improve on the Barclays integration strategy? Also, what is the optimum size of the new combined network, and which ones should we close, and where? Table 9.5 shows the results of this search for a range of different reduced network sizes, compared with the performance of the existing configuration (i.e., a strategy of 'do nothing').

The first thing we notice is that by considering rationalization alone, it is difficult to achieve the same level of performance, in terms of sales volumes and revenues with a smaller number of branches. In all cases–as was the case in the planned Barclays rationalization strategy outlined in the preceding text–the impact of a reduction in network size on revenues is negative. The reasons behind this occurrence are linked to customer deflections to competing outlets, as described in Section 9.3. The only situation in which rationalization alone is feasible (i.e,. it won't result in revenue losses), is when there is no local competition.

A more obvious area for improvement from branch rationalization strategies is in terms of network efficiency/profitability. As we can see, compared with the performance of the existing combined network, by reducing network size, efficiency is improved in all cases. It is interesting to make a 'like for like' comparison in terms of network size between the performance of the 57 outlet locations under Barclays preferred strategy, and that of the 57 'optimal' locations (Table 9.5, strategy 4). We find that, although the total size of the network remains the same, by making different closure decisions, we can improve revenues by 10% and efficiency by 11%, as compared with the Barclays network plan. In terms of the optimum configuration, we find that on the basis of maximizing revenues, and network efficiency (sales per outlet–equally weighted) rationalization strategy 8 is best. This strategy–which reduces the size of

Table 9.5 Alternative network integration strategies for Barclays/Woolwich–rationalization only

Strategy	No.	Network size	Sales	Sls/Outlet	% Change	
					Volume	Sales/Out
Existing Scenarios	na	67	48 845.75	729.04	na	na
	na	57	41 809.39	733.50	−14.41	0.61
Rationalization Scenarios	1	60	46 024.65	767.08	−5.78	5.22
	2	59	46 710.34	791.70	−4.37	8.59
	3	58	46 385.67	799.75	−5.04	9.70
	4	57	46 505.73	815.89	−4.79	11.91
	5	56	46 441.93	829.32	−4.92	13.75
	6	55	46 622.56	847.68	−4.55	16.27
	7	54	46 364.05	858.59	−5.08	17.77
	8	53	46 681.93	880.79	−4.43	20.82
	9	52	46 099.81	886.53	−5.62	21.60
	10	51	45 866.09	899.34	−6.10	23.36
	11	50	45 817.46	916.35	−6.20	25.69
	12	49	45 892.04	936.57	−6.05	28.47
	13	48	45 618.64	950.39	−6.61	30.36
	14	47	44 753.90	952.21	−8.38	30.61

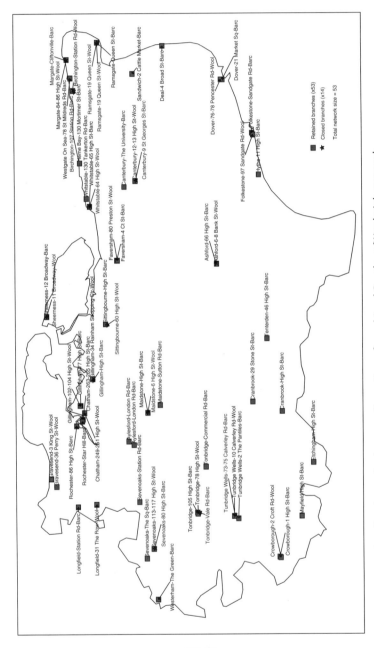

Figure 9.9 Rationalization only–best configuration of combined Barclays/Woolwich network

the network to 53-improves network efficiency by 20%, an improvement of 21% over Barclays preferred strategy. It also minimizes the negative impact on revenues with a 4% reduction in volumes, compared with the 14% reduction under Barclays strategy. The new network distribution under this strategy is shown in Figure 9.9. It can be seen that in some centres, where the combined network results in a situation in which there are now two outlets within one hundred meters of each other, both are retained. In fact, only four of the closures are the same as those under Barclays preferred strategy.

Next, let us look at relocations only. That is, let us relocate the same number of outlets to their 'ideal' locations. How many of the existing are in their optimum locations? What level of performance can we achieve through relocation? Table 9.6 shows the results of this exercise on network performance. We find that only 24 of the existing 67 outlets (36%) are located optimally. By relocating the remaining 43 outlets to new locations, we can increase revenue volumes by more than 14%. In addition, such a strategy will-compared with Barclays preferred integration strategy-increase network efficiency by more than 14%. We cannot, however, achieve the same level of efficiency as the optimum network containing 53 outlets (identified in the previous search) without closures. Furthermore, such a strategy is, perhaps, unfeasible in reality because of high one-off costs incurred in relocation. Next, therefore, let us combine the two search procedures and look at rationalizations and relocations simultaneously. Can we achieve balance between efficiency gains and revenue growth?

Table 9.7 shows the results of the optimization process in which we consider both rationalization and relocations. As we can see, it is possible to achieve both cost savings and efficiency gains, while at the same time, increasing revenues. Thirteen of the twenty network configuration strategies shown in the table achieve this balance. We find that the best configuration-again based on maximizing revenues and efficiency (equally weighted)-is strategy 7. This strategy, in which 27 of the existing branches are retained, 27 are relocated, and 13 are closed, improves network efficiency (and subsequently network profitability) by 30%. In addition, the strategy also achieves a 5% improvement in revenue volumes, compared with the existing configuration. This is a 30% improvement on Barclays preferred integration strategy! The distribution of this network configuration is shown in Figure 9.10.

Finally, it is interesting to examine whether there is a need to rationalize the network at all. Is there scope for network expansion? Can we both relocate existing outlets and open additional outlets? Table 9.8 shows the results of this exercise, compared with both Barclays strategy and the existing configuration. As we can see, there is considerable scope for expansion. By reorganizing the existing network, and opening as many as 19 additional outlets, it is possible to greatly increase revenues and, at the same time,

Table 9.6 Alternative network integration strategies for Barclays/Woolwich–relocation only

Strategy	No.	Network size	Sales	Sls/Outlet	% Change Volume	% Change Sales/Out
Existing Scenarios	na	67	48 845.75	729.04	na	na
	na	57	41 809.39	733.50	−14.41	0.61
Relocations	na	67	56 055.49	836.65	14.76	14.76

Table 9.7 Alternative network integration strategies for Barclays/Woolwich–rationalization & relocation

Strategy	No.	Network size	Sales	Sls/Outlet	% Change Volume	% Change Sales/Out
Existing Scenarios	na	67	48 845.75	729.04	na	na
	na	57	41 809.39	733.50	−14.41	0.61
Rationalization	1	60	53 764.90	896.08	10.07	22.91
& relocation	2	59	52 823.76	895.32	8.14	22.81
	3	58	52 506.93	905.29	7.50	24.18
	4	57	52 784.20	926.04	8.06	27.02
	5	56	49 979.55	892.49	2.32	22.42
	6	55	51 018.89	927.62	4.45	27.24
	7	54	51 599.74	955.55	5.64	31.07
	8	53	50 274.10	948.57	2.92	30.11
	9	52	49 736.44	956.47	1.82	31.20
	10	51	49 822.73	976.92	2.00	34.00
	11	50	49 647.36	992.95	1.64	36.20
	12	49	49 165.15	1003.37	0.65	37.63
	13	48	48 355.77	1007.41	−1.00	38.18
	14	47	47 599.74	1012.76	−2.55	38.92
	15	46	47 345.55	1029.25	−3.07	41.18
	16	45	46 755.06	1039.00	−4.28	42.52
	17	44	46 404.22	1054.64	−5.00	44.66
	18	43	45 435.34	1056.64	−6.98	44.94
	19	42	45 485.60	1082.99	−6.88	48.55
	20	41	44 632.58	1088.60	−8.63	49.32

increase network efficiency (and subsequently profitability). Again, such expansionary strategies may be unfeasible in practice, given the large-scale capital investment involved in relocation and new outlet development. A more accurate reflection would need to consider such factors. In addition, given the rise in non-physical delivery in retail banking, such analyses need to be placed within a more dynamic framework, incorporating forecasts of future channel usage, and possibly competitor reactions. Such an analysis is beyond the scope of the present paper. Nevertheless, it is interesting to observe that all the above expansion strategies offer a significant improvement on Barclays proposals, both in terms of revenue growth and network efficiency. We, therefore, conclude that Barclays preferred integration strategy is not the optimum. In fact, as we have seen, this is far from the case. We argue that the basis of their strategy–close all outlets within one hundred metres of an existing branch–is far too simplistic. As covered above, however, such a strategy is frequently pursued in practice. These strategies are undertaken with no consideration of the small-scale variations in consumer behaviour and competitive structures that exist within local markets. Without such consideration–without detailed local market 'intelligence'–we would assert that they cannot devise accurate network integration strategies.

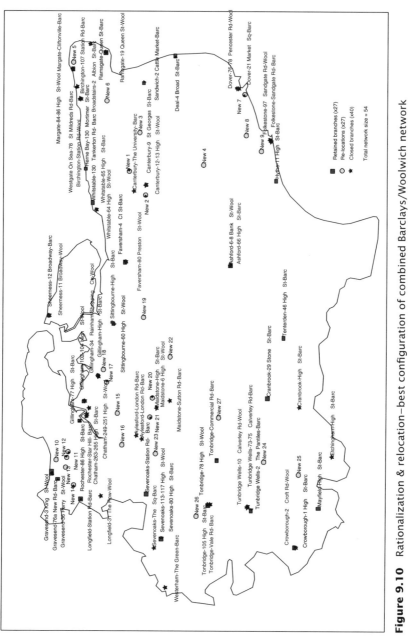

Figure 9.10 Rationalization & relocation–best configuration of combined Barclays/Woolwich network

Table 9.8 Alternative network integration strategies for Barclays/Woolwich–relocation and new openings

Strategy	No.	Network size	Sales	Sls/Outlet	% Change Volume	% Change Sales/Out
Existing Scenarios	na	67	48 845.75	729.04	na	na
	na	57	41 809.39	733.50	−14.41	0.61
Relocation	1	68	57 440.28	844.71	17.60	15.87
& new openings	2	69	57 769.28	837.24	18.27	14.84
	3	70	57 890.03	827.00	18.52	13.44
	4	71	58 097.02	818.27	18.94	12.24
	5	72	58 922.76	818.37	20.63	12.25
	6	73	58 970.77	807.82	20.73	10.81
	7	74	59 130.29	799.06	21.06	9.60
	8	75	59 444.23	792.59	21.70	8.72
	9	76	59 633.46	784.65	22.09	7.63
	10	77	60 280.25	782.86	23.41	7.38
	11	78	60 560.74	776.42	23.98	6.50
	12	79	61 050.27	772.79	24.99	6.00
	13	80	61 550.25	769.38	26.01	5.53
	14	81	61 880.74	763.96	26.69	4.79
	15	82	62 020.03	756.34	26.97	3.74
	16	83	62 499.44	753.01	27.95	3.29
	17	84	62 874.32	748.50	28.72	2.67
	18	85	63 087.68	742.21	29.16	1.81
	19	86	63 440.54	737.68	29.88	1.19

9.5 Retail network optimization

Historically, retail planning may be characterized as both reactive and subjective. It has been subjective in the sense that decisions are often made by nominated individuals on the basis of experience and 'gut feel' (see Chapter 7). The subjective element of the process may be likened, most closely, to something like property valuation. Just as an estate agent or valuer will use the knowledge of previous sale values, local market conditions, and a detailed survey of an individual property, so a retail market analyst will use detailed investigations of a local market, the knowledge, and experience of similar stores and towns to make a judgement. This is, without doubt, a skilled process. The reactive component arises in the sense that assessment typically follows availability of a suitable property. Once a retailer becomes aware that space may be available in a town that lacks a suitable presence, then an analyst may be instructed to perform an assessment of the location.

There are many organizations in which this style of planning continues to be adopted, albeit usually supported by empirical data regarding the local market, such as demographics and competitive intelligence. Although the views of an experienced analyst may carry much weight, the drawbacks to this approach are well rehearsed. Endless field observation is generally involved, which is expensive and time-consuming. Local markets are becoming increasing niched on the demand-side and competitive on the supply-side, making them increasingly complex. It is difficult to make large numbers of assessments, or to make judgements about relatively subtle adjustments to formats. For example, what happens if we sacrifice a bit less parking for a little more store space?

The use of models, of whatever type (see Chapter 7), introduces objectivity to this process. Individual appraisals can be conducted with much greater rapidity. Provided the same assumptions are made, different analysts will generate consistent conclusions.

Systematic evaluation of network potential within a 'representation planning' framework introduces a further and more strategic dimension to the modelling process. Rather than focusing on individual locations, it is now possible to look at strengths and weaknesses in a network and to target development opportunities: for example, 'go and find a site with at least 60 000 square feet to the west of Bradford.' In the previous sections, we have seen how a spatial modelling capability can be applied to determine an ideal configuration of locations on a network. The essence of the problem, therefore, is to select N locations from a possible M, where the number of potential locations (M) is very large relative to the number to be selected (N). Here, we consider a variant of the problem that may be referred to as *format optimization*. In this application, the locations are determined. The variable is the layout and product mix (format) to be adopted at each location.

A good example of the requirements for optimization of formats is provided by petrol forecourts. In recent years, downstream petroleum retailers in Europe have found their fuel margins to be increasingly squeezed, especially through the impact of progressively higher taxation and new market entrants, notably supermarkets. Many have responded by trying to maximize the value of the real estate provided by the petrol forecourt, partly through the introduction of ancillary services such as car washes, ATMs, and even Quick Service Restaurants in some cases; but, most importantly, through a more disciplined approach to forecourt shops. Where petroleum retailers have recognized that they lack the specialist skills of grocery retailers or newsagents, they have sought to build up those skills, often through partnership (for example, BP with Safeway and Esso with Tesco).

Petrol stations are also interesting in that, typically, they serve a variety of markets, from the highly transient trunk road sites and motorway service stations, to local stations within small towns. At the extremes, these locations provide highly differentiated retail opportunities. A typical problem for a petrol retailer would be to develop a variety of formats at different location types, for example:

- CTN–a format based on a conventional newsagent (Confectionery-Tobacco-News), to be targeted at transient urban sites with well-marked business peaks in the morning and evening rush hours;
- Pit stop–a snack food format with hot drinks, microwave ovens, fresh sandwiches and so forth focused to motorway and trunk road sites with a long-distance clientele.
- Fresh Foods–a format geared to a local population, with a high proportion of walk-in (non-fuel) trade. Likely to be effective in areas with poor competitor provision of grocers or supermarkets.

The representation-planning requirement is to identify the appropriate format for each site in the network, in relation to the customer profile and competitor profile of each site. This is a relatively simple task, compared with the IRP discussed previously, because there is a low level of interdependence between different forecourts. The format that is adopted at a particular BP station is unlikely to be materially affected by the format adopted at neighbouring BP stations. Some example assignments for sites in the Leeds area are shown in Table 9.9, in which the data relating to individual locations is real, but the brands have been anonymized. In general, the table indicates that many of the lower income locations, such as Burmantofts and Crossgates, tend to favour a convenience branding, while the more affluent suburbs are more appropriate for fresh foods. Pit stops are most well suited to trunk roads and high volume urban carriageways.

Reality is, however, more complicated than Table 9.9 suggests, because actually no two sites are exactly alike. To look at this a different way, any site has some combination of CTN, Pit Stop or Fresh Food within its make-up. This view is represented in Figure 9.11, which shows the mix at each site relative to a triangle in which the vertices represent the three major formats. Each of the numbered circles refers to the sites shown previously in Table 9.9. The locations close to the vertices show sites that are well characterized by the simple labels of CTN (e.g. Blue Square, Seacroft–number 23), Pit Stop (e.g. Red Circle, Guiseley–number 27), and Fresh Foods (e.g. Gold Star, Headingley–number 21). Locations

Table 9.9 Simple format optimization for Leeds sites

Brand	Suburb	Town	Type
Green Triangle	Beeston	Leeds	C
Blue Square	Burmantofts	Leeds	C
Red Circle	Oulton	Leeds	C
Red Circle	Crossgates	Leeds	C
Green Triangle	Chapel Allerton	Leeds	C
Green Triangle	Shadwell	Leeds	C
Blue Square	Seacroft	Leeds	C
Black Spot	Kirkstall	Leeds	C
Blue Square	Meanwood	Leeds	C
Green Triangle	Rothwell	Leeds	F
Gold Star	Pudsey	Leeds	F
Black Spot	Pudsey	Leeds	F
Gold Star	Harehills	Leeds	F
Green Triangle	Rothwell	Leeds	F
Black Spot	Bramley	Leeds	F
Gold Star	Headingley	Leeds	F
Gold Star	Gledhow	Leeds	F
Blue Square	Redhill	Leeds	P
Gold Star	Headingley	Leeds	P
Green Triangle	Headingley	Leeds	P
Red Circle	Horsforth	Leeds	P
Gold Star	Yeadon	Leeds	P
Blue Square	Moortown	Leeds	P
Red Circle	Moortown	Leeds	P
Blue Square	Roundhay	Leeds	P
Red Circle	Hunslet	Leeds	P
Red Circle	Guiseley	Leeds	P
Black Spot	Armley	Leeds	P
Black Spot	Killingbeck	Leeds	P
Gold Star	Beeston	Leeds	P

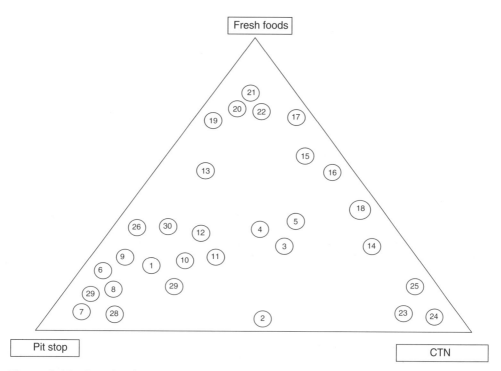

Figure 9.11 Complex format optimization for Leeds sites

such as Red Circle, Crossgates (number 4), and Green Triangle, Chapel Allerton (number 5), look much harder to classify.

A logical next step would be to refine the optimization model to suggest an exact format for each location on a continuous scale, of which the three basic formats are the most extreme examples. This concept is illustrated in Table 9.10. Here, we see that the classic CTN format has a concentration on the three core products of cigarettes, confectionery and newspapers & magazines, with some chilled foods and staples (e.g. tinned products, cereals). The Pit Stop format has all the elements of a CTN in smaller measure, but with a major emphasis on snack foods. Fresh Foods has a major focus on chilled foods and staple groceries, and a distinctive emphasis on fresh produce. However, a more extensive variety of formats that might be appropriate at locations around the network are also shown. For example, the Type 6 format is 'halfway' between the Fresh Foods and Pitstop formats. This layout might be appropriate for a location such as Black Spot, Headingley (number 18).

In conceptual terms, the adoption of continuously optimized formats, as described in the preceding text, is not difficult, although, in practice, many established retailers may find the logistics of supplying a variable product mix to different sites more difficult than a standard mix. With the progression of technologies for automated control of inventories and stocking, however, this is becoming much less of an issue than would have been the

Table 9.10 Continuous site formats

	CTN	Fresh foods	Pitstop	Type 4	Type 5	Type 6	Type 7	Type 8	Type 9	Type 10
Cigarettes	30	10	20	20	20	25	15	15	20	25
Confectionery	30	10	20	20	20	25	15	15	20	25
News	30	10	20	20	20	25	15	15	20	25
Chilled foods	0	20	0	5	10	0	10	15	5	5
Staples	10	20	0	15	15	5	10	15	5	10
Snack foods	0	0	40	10	0	20	20	5	25	5
Fresh foods	0	30	0	10	15	0	15	20	5	5

Note: CTN-Confectionery-Tobacco-News

case ten years ago. Retailers have always worked with the concept of seasonal variations in stocking (e.g. winter clothes and summer clothes; salads; Christmas cards and gifts) and are beginning to experiment with daily variations in format. For example, the space devoted to sandwiches at the middle of the day might be replaced by cooked chilled meals in the evening. In the opinion of the authors, the ability of retailers to optimize their product mix to extract maximum value from local markets will become a major interest for retailers in the none-too-distant future.

9.6 Conclusion

In this chapter, we have looked at the problem of network planning and introduced the 'Idealized Representation Plan' as a device for solving problems of this type. We have discussed examples from the Automotive, Financial Services and Retail sectors. The problems that can be addressed range from defining networks when entering a new geographical market, achieving effective integration between two networks following an acquisition, and providing the right shop formats at the right locations.

Typically, these problems are both mathematically and computationally complex, so that obtaining effective solutions requires a blend of analytical imagination and computational brute force. However, given the ready availability of ever-increasing quantities of computing power, the ability to implement procedures of this type can only get easier. Hence, the central message of the chapter is that extra computational power can provide the means to solve new and interesting classes of the locational problem. A similar theme is expounded in Chapter 11, Section 11.3.

DIRECT MARKETING AND DISTRIBUTION CHANNEL MANAGEMENT

10.1 Introduction

In this chapter, we will consider marketing approaches through which products are sold directly to the customer, and also distribution processes that require a complex mix of different channels to market. We begin, in Section 10.2, with a view of geodemographics. An overview of geodemographics has been presented earlier in Chapter 7, Section 7.4. Here, we focus on the suitability of geodemographics as a basis for direct marketing. In view of some of the weaknesses of geodemographics, one approach is to try to enhance these systems using more powerful methodologies. One such approach is fuzzy modelling, and this is considered in Section 10.3. However, a more intractable weakness of geodemographic approaches is their dependence on area level data. The movement towards the use of individual level, or 'lifestyles' data, is described in Section 10.4 of the chapter. The development of sophisticated 'data mining' technologies to large customer data sets is then reviewed. Finally, in Section 10.6, we present a view of the kind of modelling capability that may support distribution through multiple channels.

10.2 Geodemographics

10.2.1 Introduction

Three topics of interest relating to geodemographics will be considered below – the application of geodemographics to direct marketing, the strengths of the approach, and its weaknesses.

The technique of geodemographic analysis, involving the application of a classification system to small area data (see Chapter 7, Section 7.4), has a number of applications within marketing and retail analysis, besides direct marketing. Thus, CACI (1993) have identified nine application areas for their ACORN (A classification of Residential Neighbourhoods) product:

- Site analysis
- Sales planning

- Planning for public services
- Media buying
- Database analysis
- Market research sample frames
- Direct mail
- Coding
- Door-to-door leaflet campaigns

These applications are discussed in more detail by Beaumont (1991), using a framework of four headings:

1. Branch Location Analysis: Applications to branch location analysis have already been considered in Chapter 7, although we have argued in Chapters 7 and 8 that there are often more appropriate techniques than geodemographics for this purpose.

2. Marketing Management Information Systems (MIS): When geodemographics are integrated with geographical information systems (GIS) (see Chapter 7, Section 7.4.2), we start to arrive at something like a Marketing MIS. For example, we can start to understand the characteristics of areas where market penetration is strong, and where it is weak; we can understand the demographic profile of individual branches or outlets, and target customers or format the branches, accordingly; and we can understand network performance issues as a basis for remedial strategies.

3. Credit scoring: Credit scoring applications are typically based on some kind of scorecard that is similar to the location scorecards, which we looked at in Chapter 7, Section 7.4.3. In addition to characteristics such as customer's income and existing expenditure, age, family status and so forth, the postcode is commonly used as a qualifier in determining whether a customer has sufficient credit-worthiness for financial transactions.

4. Direct Marketing: Direct marketing is by far the most significant application category for geodemographics, and to this we now turn.

10.2.2   Geodemographics and direct marketing

Direct Marketing applications are based on the idea that different kinds of products appeal to different kinds of people. For example, readers of the Sun newspaper, typically, have quite a different profile from readers of the Times, and this may apply to age, gender, social class, educational attainment and so on. On the basis that geodemographic classifications encapsulate variations in characteristics such as these, one can expect to find variable product penetrations across different geodemographic groups. Now, if an organization has a new product that it wants to market, and knows (from extensive pre-sales market research and testing) what kinds of people are likely to be attracted to the product, then it can simply target activities towards customers living in areas with a high response potential. This is important, for example, if people are to be sent promotional brochures through the post, which are expensive to produce and distribute. So, if it costs x to produce and deliver a piece of promotional material, and there is a potential return (profit) of $100x$ if a customer makes a positive response to the promotion, then it might be appropriate to

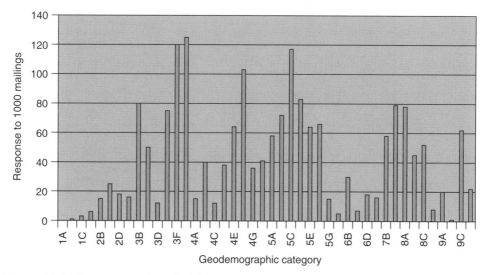

Figure 10.1   Response analysis for life insurance

focus promotions on those areas where a response rate of at least 1% is expected. Hence, this kind of marketing activity is often also known as 'response modelling'.

An example is shown in Figure 10.1, in which a mailing for life insurance products has been analyzed for different geodemographic groups. Whereas the overall response to the mailing runs at between 3 and 4% (i.e. 30 to 40 responses per 1000 mailings), these rates are much higher for groups such as 3F, 3G and 5C. These are groups with a high concentration of affluent, young to middle-aged families. Response rates are much lower for groups such as 1A, 1B, 6A and 9B. These are groups with high concentrations of young single people with little disposable income.

The responses of Figure 10.1 are shown as a 'Gains Chart' in Figure 10.2. The gains chart shows the highest levels of response that can be achieved from a given share of the population base (for more detail about gains charts, see Birkin, 1995). When the responses are highly segmented, as in Figure 10.1, then the gains chart shows that a high proportion of customers can be accessed via highly targeted campaigns. Thus, we can see in this case that half the customers are concentrated within only about 20% of the population base, whereas 50% of the population accounts for about 80% of customers.

10.2.3   Strengths of geodemographics

Geodemographics has had a robust and enduring pedigree for a number of years (e.g. Brown, 1991; Sleight 1993; Batey and Brown, 1995; Birkin, 1995 for good review material). The strengths of the technique are that it uses high-quality data; it is easy to understand and easy to apply; and it is easy to quantify and visualize the outcomes. Each of these aspects is now considered in a little more detail:

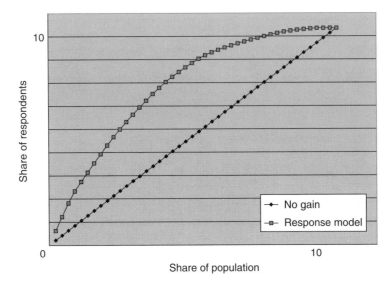

Figure 10.2 Gains chart for life insurance

1. High-quality data: A major strength of geodemographic systems is that they are based on large quantities of high-quality census data. The census has coverage that is comprehensive, both geographically and in its penetration of the population base – it is a criminal offence in the United Kingdom not to return a census form.

The census provides very detailed demographic data, for example, population by single years of age; detailed household composition – numbers of people and their interrelationship; household migration; place of birth and so on. The census also includes economic activity information not available elsewhere, such as detailed occupation and industry of employment, and journey-to-work data.

Census data is released for small geographic areas – approximately 150 000 enumeration districts (ED) in the United Kingdom 2001. For each ED, approximately 4500 cell counts were released in 1991. Thus, 150 000 districts times 4500 counts yields 675 million census cell counts.

2. Easily understood: Geodemographics is based on the idea that each small area can be assigned a (hierarchical) label. Each of these labels is, in turn, associated with a *'pen profile'* that describes the typical residents of that area. An example is shown in Table 10.1.

3. Easy to apply to customer data: The concept of address matching can be used to add geodemographic codes to any database that includes customer addresses. The database will generally need to be cleaned and updated: if postcodes are not included, they may be added from the Postcode Address File; incorrect postcodes replaced; partial codes made complete; and out-of-date postcodes updated. Then the geodemographic labels can simply be 'looked up' in an appropriate directory.

Table 10.1 Rural retirement mix

MOSAIC Group M46: RURAL RETIREMENT MIX

Whereas the Worthings and Sidmouths continue to attract a disproportionate share of the nation's retirees, an increasing number of mobile older people prefer to retire to small detached bungalows in rural areas often close to the sea or other natural attractions.

Rural Retirement Mix consists of a small segment of the population in which traditional farming populations have been joined by rural retirees and small proprietors in the tourist industry.

Such areas are characterized by high proportions of owner-occupiers, living in detached and 'named' dwellings, often in small seaside villages. Pensioners, although numerous, do not dominate the community to the same extent as in the Independent Elders. In these areas, old people are typically married and still active.

In such areas, people's interests are in gardening and local environmental activities rather than in foreign travel. Going to church and supporting local charities is more important than keeping up with latest fashions and technology. People dislike junk mail and neither need nor use credit.

4. Easy to quantify outcomes and visualize: We have already shown how to quantify outcomes using a gains chart, and in Chapter 7, Section 7.4.2, how to visualize the data using GIS systems.

10.2.4   Weaknesses of geodemographics

Unfortunately, there are also a number of problems with geodemographics. Let us consider a number of these problems in turn:

1. Ecological fallacy: The ecological fallacy can be broadly stated as the assumption that individuals within a small geographical area share the same characteristics that are common to the area. For example, if an area is characterized as 'middle-class families', then each of its residents is middle class and living in a family.

Consider the example shown in Table 10.2, which shows the profiles of the main Superprofiles clusters against selected census variables.

Table 10.2   Census variables by superprofiles cluster (A–J)

Census variable	UK average	A	B	C	D	E	F	G	H	I	J
Persons aged 0–4	5.8	73	74	139	87	78	116	138	99	75	138
Persons aged 5–14	14.0	95	60	125	97	86	115	106	87	92	138
Persons aged 15–24	15.6	98	111	86	90	82	95	121	101	101	121
Persons aged 25–44	26.1	91	115	135	95	90	116	100	97	80	97
Persons aged 45–64	22.7	114	92	78	107	108	91	86	96	119	84
Persons aged 65–74	9.7	106	109	51	111	140	73	77	118	120	61
Persons aged 75+	6.0	119	138	45	111	147	64	74	121	114	49
Single worker households	7.8	87	267	64	85	76	66	141	118	82	65
Married couple households	44.8	100	70	136	105	105	121	78	94	87	94

Extreme and obvious examples of the ecological fallacy can be seen in geodemographic categories such as Cluster B, 'Metro Singles'. Within this cluster, single workers are indexed at 267, which means that they are more than twice as likely to be found in these areas than in the population as a whole. However, looking at Type B areas, we can see that only 20% of people living in these areas are single workers; whereas more than 30% of households are married couples.

There are elements of the ecological fallacy even in the most robust clusters. For example, Superprofiles Cluster C is labelled as 'Young Married Suburbia', yet 25% of the population is aged 45 or over.

The effects of ecological fallacy can also be seen if we look at the problem from the perspective of a commercial organization seeking to target its customers. For example, suppose Mothercare is looking for new outlet locations in UK retail centres and wishes to consider the quality of the retail catchment. To do this, the retailer might select an appropriate group of geodemographic codes, such as Superprofiles C, G and J. However, once again, although the concentration of target families is high in these areas, they only account for a minority of potential customers. The retailer would probably do better to take a straightforward count of young children by small area from the census of population.

2. *Attitudes and values:* A major drawback of census data is that it takes no account of the pastimes, attitudes, values and preferences of the population, which can ultimately expect to be revealed in customer purchasing behaviour. For example, we cannot find out from the census whether golf is more popular in the north of England than in the south; or whether people living in Essex really like to drive fast cars. It is towards these kinds of issues that 'lifestyle' questionnaires are most directly focused.

3. *Static data:* Although the census is comprehensive, it is also infrequent. In the United Kingdom, the census has traditionally been decennial, that is, it is conducted once every ten years every time the year ends in 'one'. In 1966, there was an interim quinquennial census, but this was not maintained because of the cost of the exercise. Thus, the census provides a snapshot of the population at a point in time, with each snapshot taking rather a long time to come around. This problem is compounded by the fact that after the census is conducted it typically takes around three years to get the data processed and released. So, by the time 2001 census data is released, data from the previous census will be fully thirteen years out-of-date.

4. *General-purpose tool:* Although census-based systems provide an excellent overview of the population, they fail to provide detailed information about specific activities, and especially about commercial markets. It is increasingly necessary for organizations to consider whether what they know about their own customers (e.g. from store loyalty cards or transaction histories) is of comparable or greater value than general-purpose classifications derived from the census.

10.3   Enhancement of Geodemographics

10.3.1   Fuzzy geodemographics

The classification approach adopted within conventional geodemographic systems presents an assignment of geodemographic place types to each enumeration district within the database. Thus, we have a one-to-one mapping of EDs to cluster types. This is sometimes referred to as a *crisp* technology. Neither of the following is allowed:

● It is not possible to belong to more than one cluster simultaneously.
● It is not possible to be 'close to' a cluster or 'nearly in' a cluster.

To an extent, therefore, the traditional targeting process adopts boundaries that are arbitrary and somewhat artificial. We can try to make this more flexible using the concept of fuzzy geodemographics.

The concept of fuzziness is borrowed from fuzzy logic/ fuzzy mathematics (e.g. Zadeh, 1965). It provides a way of looking at uncertainty. There are two types of uncertainty that are relevant to geodemographics:

1. Fuzziness in the classification space
2. Fuzziness in geographical space

1. Fuzziness in the classification space: Consider the 'Butterfly Example' shown in Figure 10.3. The objective is to segment the circular data points into two classes. Clearly, there is a problem in this situation since the middle point might equally be assigned to either class. Equally, however, it may be observed that the data points labelled 'A', which are close to the centre, are less clearly assigned to their respective classes than the wing tips, labelled B. These uncertainties may be reflected using a fuzzy classification process, that produces a membership matrix, such as that shown in the following text. The middle point of our butterfly example is something like Data Case 2 being split equally between a number of clusters. The wing tips are more like Data Case 4 – they are, clearly, associated with a particular cluster. The A-points are analogous to Data Case 3 – they are probably assigned to the right cluster, but there is less certainty about this.

The application of a fuzzy classification system would rely on the selection of cases within particular clusters at varying degrees of certainty. For example, if EDs were classified in this way, then we might select all EDs with a 75% membership (say) of cluster 1 rather than selecting Eds that are 'known' to belong to cluster 1 and no other. If this threshold is reduced to levels below 50%, then it is clearly possible to select the same case in association with different clusters. Alternatively, selections could be made on the basis of multiple membership, for example, at least 30% membership of both clusters 1 and 2.

2. Fuzziness in geographic space: Fuzzy geographies are interesting because the whole philosophy of geodemographics is based on the existence of 'neighbourhood effects',

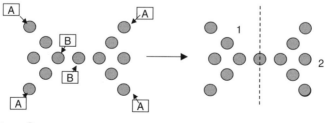

Figure 10.3 Butterfly

with similar people coming together within local areas. The problem with conventional geodemographic classifications is that these neighbourhood effects are forced to operate at an enumeration district level. As Openshaw *et al.* (1995) argue:

> Geographers in the GIS era really should be able to do better than this and regard spatial neighbourhood effects in an elastic fashion rather than as a discrete geography space

For a hypothetical example, suppose area *X* has been classified using a fuzzy process, with membership as follows:

Cluster 6	40%
Cluster 9	40%
Cluster 12	20%

Area *X* has one neighbour (Area *Y*) with the following membership:

Cluster 6	10%
Cluster 9	80%
Cluster 12	10%

We might conclude that the favoured assignment of Area *X* is actually to cluster 9, because of its neighbour's characteristics.

A more substantial example is provided by Openshaw *et al.* (1995), using an analysis of fuzziness in the assignment of EDs to clusters using a neural net based classifier. Each ED is assigned to a particular cluster using a procedure of the usual crisp type. However, the relationship to other EDs is also considered in relation to both geographic distance and cluster similarity distance. An example is shown for a particular cluster – 'well-off metro singles' – in Table 10.3. The first column of the table shows the distribution of nearest neighbour geographic distances for EDs in the selected cluster. For example, 1185 EDs of

Table 10.3 Fuzzy similarity between geodemographic clusters

Cluster No. 19: Well-off metro singles   **Members = 3849**

Cluster similarity distance

Geographic distance	0	0.25	0.5	0.75	1	1.5	2	3
0	4	1	1	0	1	0	0	0
100	1641	883	477	177	146	85	15	3
200	1185	1156	1068	637	429	328	129	51
300	309	701	971	800	565	579	270	163
400	158	505	689	781	544	646	331	245
500	91	313	577	683	546	675	343	307
750	136	449	1128	1316	1151	1510	960	977
1000	58	295	689	884	906	1341	937	1055
2000	116	395	1113	1738	1918	3252	2921	3711
3000	47	161	393	629	812	1769	1690	2823
3000+	104	416	1415	2590	3532	9139	11076	23331

cluster 19 lie between 100 and 200 metres from a similar ED. Readings across the table show similar information for clusters that are further away in the classification space. This table shows clear evidence of neighbourhood effects, since the largest numbers of high similarity clusters are found close by (cells in top left hand corner), whereas the most dissimilar clusters tend to be geographically distant (cells in bottom right hand corner).

Openshaw (1998) has a used a fuzzy k-means approach to develop a fuzzy version of MOSAIC. The approach was tested using a telephone response data set provided by a major bank for around 200 000 postcodes. Gains Charts were created for each classification to find their effectiveness in targeting response from particular sub-groups of the population. The results are summarized in Figure 10.4. At best, it may be concluded that the results from this experiment are inconclusive. It appears that the process of fuzzy response modelling may require further refinement before its superiority to crisp technologies can be definitively proven (see also Feng and Flowerdew 1997, See and Openshaw, 2001).

10.3.2 Neighbourhood lifestyle segmentation

An important recent development is the attempt to explicitly combine and incorporate lifestyle and other data sources within geodemographic products. A good example of this is the PRIZM product from Claritas. This should be considered as a geodemographic system because what it seeks to do is to assign distinct cluster types to geographic areas. However, the technology is distinctive because:

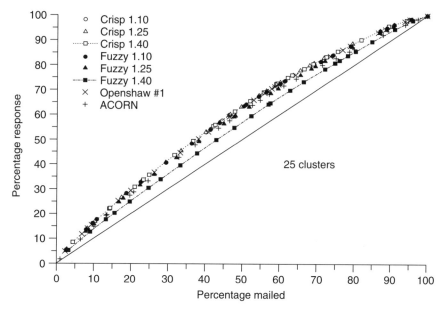

Figure 10.4 Gains chart

- The neighbourhoods are defined using unit postcodes rather than enumeration districts
- In addition to census data, the classification incorporates a combination of data sources

  - Lifestyle data from both CMT ('Computerized Marketing Technologies') and NDL ('National Demographics and Lifestyles') (Claritas, CMT and NDL are now all under common ownership)
  - Company Directors (from Dun & Bradstreet)
  - Share ownership (from NDL)
  - Behaviourbank (from CMT)
  - Unemployment data (from Dept of Employment)
  - Birth and death rates (from the Office for National Statistics – ONS)
  - Housing name data (from the Postcode Address File)

72 PRIZM clusters have been identified, although these are in turn mapped onto 4 lifestage and 5 income groups, as follows:

Lifestage
1. Starting Out
2. Nursery families
3. Established families
4. Empty nests

Income quintile
1. Lowest household incomes
2. Moderately low household incomes
3. Intermediate incomes
4. Moderately high household incomes
5. Highest household incomes

All of this yields a slightly complex structure in which each Prizm cluster has a four-character code, such as B218. This means PRIZM cluster 18 (last two digits), which falls within Lifestage B (Nursery families) and income group 2 (low to moderate). B218 is labelled as 'Fresh Air Futures'.

A sample of clusters from the PRIZM classification is shown in Table 10.4 below:

Table 10.4 Sample clusters from the PRIZM classification

Cluster label	PRIZM code	Income ranking	Age ranking	% households
Cosy couples	A210	35	24	0.78
Small town and tots	B220	31	6	2.67
Battling families	B530	72	32	1.77
Gentlemen farmers	C240	24	43	0.18
Down beat uplands	C350	45	40	0.97
Jams and geraniums	D260	28	64	1.82
Ever hopefuls	D570	66	62	2.79

PRIZM solutions also exist for the United States, France and Republic of Ireland (Claritas, 2001). In developing the PRIZM classification, the following methods were adopted (Claritas, 1996):

1. Reweighting. Census data has been used to weight the lifestyle variables to remove 'all known biases' in the data.

2. Uncertainty. A method of imputation with certain 'fuzzy' overtones has been used to populate postcodes with little or no lifestyle data. In this situation, the response frequency is enhanced by considering adjacent geographic neighbourhoods.

3. Multiple clusters. A number of different cluster solutions totalling 29 were developed and evaluated to assess which system provided the highest levels of discrimination.

10.3.3 Other developments

As we have seen, the original geodemographic providers – notably CACI and CCN-Experian – are threatened in a number of ways by the advent of new technologies and new databases into their traditional domains. At least three possible responses are available to these providers:

1. *Improvement to the basic classification:* This could be achieved by the inclusion of a wider range of data within the base classifications, and/or by the use of better methods in the classification process. There is some evidence that Experian's MOSAIC product now includes a broader range of data than previously. For example, Thrall *et al.* (2001) suggest that 'to produce Mosaic, Experian integrated information linked to telephone numbers, property data and direct consumer responses along with other geographic-specific data.' While according to the University of Edinburgh (2001), MOSAIC includes data from the electoral roll, commercial credit information and 'other marketing data'. On the other hand, there is little evidence from the CACI homepage that ACORN has been extended to include non-census data.

Regarding the technology in their creation, CACI state simply that ACORN has been built 'using multivariate statistical methods' (CACI, 2001), whereas 'Mosaic's developers used iterative cluster analysis to establish 12 major core group and 62 subgroup classifications' (Thrall *et al.*, 2001). There seems to be little evidence of breakthrough technologies here.

2. *Derived classifications:* Many new systems are now being created through the integration of core geodemographic systems, with one or more other classifications. For example, Experian have created both 'Financial Strategy Segments' and 'Grocery Mosaic' through the integration of MOSAIC with 'Pixel' (Experian, 2001). Pixel is described as follows: "Pixel is a high-resolution segmentation system that assigns every adult in Great Britain to one of over 6000 categories. It combines seven known pieces of information on every consumer (Gender, Age, Length of Residency, Company Directorships, Shareholdings, Property Type, Household Composition) to arrive at a unique Pixel code." A similar strategy has been adopted by CACI in the creation of products such as Financial*ACORN and Investor*ACORN.

3. *Use of geodemographics in parallel with other segmentation approaches:* Another strategy, which has been demonstrated (CACI, 2001), is simply to profile a customer database against a whole variety of explanatory variables that may be pulled from geodemographic

systems (i.e. ACORN), from derived products (such as Financial*ACORN), or from Lifestyles data. In this way, it appears that the benefits of using extended data sets are being sought, without undermining the position of the flagship geodemographic product.

10.4 Lifestyle Databases

10.4.1 A world of missing databases

Expansion of databases has, like computer power, been exponential over the last 20 years. Data can now be assembled from all kinds of sources other than the census, for example:

- Shopping destinations, for example, from customer loyalty cards ('storecards'). From this source, retailers may also be able to infer the frequency of visits, customer value, spending habits and product preferences.
- Spending patterns – from credit card data or analysis of transaction patterns by current account.
- Who you talk to – from telephone records, or 'cookies', which can trace internet activity on your computer.
- Health data – includes information about births and deaths by small area (from Registry Office data) and migration (from National Health Service Central Register). Also includes data about morbidity (i.e. patterns of illness) from GP prescribing data, hospital referrals, diagnoses and treatment patterns.
- Crime – from police databases. When and where a crime was committed; nature of the offence.
- Remote sensing – provides regular and detailed images of the earth's surface from satellite data.
- Digital maps – for example, companies like Yellowpoint (a joint venture between Yellow Pages and the Ordnance Survey) can now provide street and buildings level map data, connected to the occupancy of those properties, to something approaching 1 m accuracy.
- Company data – for example, Companies House provides information about all registered companies in the United Kingdom, including Head Office, nature of business and latest accounts.

It is debatable whether the ability to extract information from data (i.e. to provide interpretation or insights leading to action) has kept pace. Current systems may be referred to as data rich, information poor; although the data is plentiful, what it all means is less clear. This is relevant to geodemographics because this is a 20-year-old technology whose power comes from a translation of complex data to simple and robust, but imperfect information (cf. Section 10.2.4).

Of course, other data sources are also routinely included within geodemographic systems – for example, County Court Judgements (CCJs), registers of company directors and share ownership and electoral roll data. The question is whether some of these other data sources could, perhaps, be incorporated within geodemographic systems, or within other database marketing strategies. If we can provide better representations, for example, lifestyle data, or draw more powerful conclusions (perhaps via modelling – see Section 10.6; or data mining (Section 10.5)), then geodemographics is likely to become redundant.

10.4.2 Lifestyle databases – background

Lifestyle databases are built from customer responses to questionnaires. The question-naires, typically, include a section on basic demographics (age, address, family and occupation) and, unlike the census, will usually feature a question relating to income. This is combined with a section on the respondent's hobbies and interests, including the papers they read or TV programmes they watch, the cars they drive and the sports they play; whether they are interested in gardening, reading, music or the arts and so on.

Four of the major lifestyles companies in the United Kingdom are NDL International (National Demographics and Lifestyles), CMT (Customer Marketing Technologies), ICD (International Communications and Data) and CSL (Consumer Surveys Limited). Three of these companies (ICD, CMT, CSL) distribute their questionnaires through door-to-door delivery, or as inserts in high circulation magazines such as the Sunday Times colour supplement. Customers are incentivized to respond by the offer of grocery coupons or entry into a free prize draw. The questionnaires are usually disguized by some kind of branding that sounds as though it might be socially useful, for example, the Facts of Living Survey (ICD) or National Shoppers Survey (CMT).

NDL uses questionnaires as part of the product registration card, which comes packed with new consumer durables such as electrical goods. Customers are also incentivized to respond by the need to register the product to be covered by the manufacturer's warranty. By 1993, NDL had obtained information on 13 million customers, or one in three UK households (Morris, 1993). CMT has 7 million individuals and 4 million households at this time.

10.4.3 Applications of lifestyle data

The basic application of lifestyles data is for customer profiling and response modelling. A list of customers is provided to the database holder and the two databases are matched. The characteristics of the customers common to both databases may be used to provide lifestyle profiles of the customer list. CMT claim that this process, typically, allows them to match 10–20% of the customer file. Morris (1993) quotes a more precise figure of 12% for NDL, and states that a file of only 5000 customers is fit for this purpose.

Customer profiling might be used to support sales promotion and product launch in the motor vehicles industry in the following ways:

1. Within a database such as The Lifestyle Selector, individual customers can be identified according to the make and model of the vehicle they drive – privately or company-owned, and its year of registration. Thus, for example, drivers of three-year old Ford Mondeos could be targeted directly in order to initiate a switch to Volkswagen Golf.

2. Customer Profile Analysis can be used to provide a picture of the drivers of different models. This information might be used in product development, for example, if there is a high incidence of outdoor interests (e.g. hunting, fishing, horses) in a particular customer group, then four-wheel drive may be a fruitful product extension.

3. Customer Profile matching can be used to find potential recruits who have similar characteristics to existing customers.

4. Competitors customers can be analyzed in the same way, to provide a basis for market growth through customer churn and conquesting.

5. Support may be provided to specific dealers by identifying prospects within the geographical catchment area of each dealer.

In a promotional leaflet entitled 'Targeting for the Car Industry', NDL (1993) provide the following illustrations:

- For the launch of a new high performance hatchback, the manufacturer wants to support a national TV campaign using direct mail. The Lifestyle Selector can be used to target owners of competing models (e.g. Peugeot 205, Golf Gti, Astra GTE) aged between 2 and 4 years.
- A dealer is planning the promotion of a specific range of pre-owned vehicles. The dealer can identify and mail high potential prospects (against the profile of previous buyers) within fifteen minutes drive time of the car showroom.
- A manufacturer wants to use direct mail to accelerate sales of a mid-range family saloon. They can target owners of 2- to 5-year-old Ford Escorts and GM-Vauxhall Novas and also refine their selection by identifying owners who share similar characteristics to the existing customer base.

The following benefits are claimed by the vendors:

- Provides access to millions of car owners
- Prospects can be found within the 'buying window' (e.g. owners with three-year-old vehicles are frequently looking to change)
- Allows selections by make, model and age of current vehicle
- Identifies prospects who share the characteristics of existing customers
- Provides selections that allow customers of the competition to be targeted
- Gives valuable insights into the marketplace
- Has sufficient flexibility to support both local and national campaigns

Two other lifestyles applications are considered by Morris (1993):

Affinity marketing refers to the process of targeting people with similar interests to current customers. He describes a hypothetical example, thus: 'The customer profile might reveal that a large percentage of the buyers of a specific product go skiing. In this instance, a marketer might decide to expand the mailing by using lists bought from ski magazines or the makers of ski equipment, and mailing to the people on the lifestyle database who had expressed an interest in skiing' (Morris, 1993, p27). Against enhancing lists, Morris (1993, p27) states simply that: 'targeting can be further refined by screening out individuals who do not fit certain criteria. This could be based on credit-worthiness, age or income, for example.'

A fourth and final application is lifestyle segmentation that works on a similar principle to geodemographics, except that the segmentation is applied to individual customers

Table 10.5 Persona types

Young affluentials;
Golf clubs and volvos;
Bon Viveurs;
Tradition and charity;
Achievers;
Trinkets and treasures;
New teachers;
Safe and sensible;
Health and humanities;
Craftsmen and homemakers;
Cultural travellers;
Carry on camping;
Wildlife trustees;
Field and steam;
Crisps and videos;
Fads and fashions;
Home and garden;
Pubs, pools and bingo;
Survivors;
Retired villagers

rather than to small areas. Individuals may be categorized according to their demographics, tastes and interests. If this process is successful, then the categories into which people fall will match their spending and consumption profiles. CMT describes this as 'Behaviour-graphics' that is embedded into a system called *Persona*. They consider the philosophy of Persona to be 'you are what you do', in contrast with geodemographics, 'you are where you live'. Within Persona, each household in the 'BehaviourBank' is allocated a unique type, with no overlaps or multiple matches. There are twenty Persona types, as shown in Table 10.5.

10.4.4 Evaluation of lifestyles data

Figures 10.5, 10.6 and 10.7 show a comparison between a lifestyles database and 1991 census Sample of Anonymized Records (SAR), using criteria of age, car ownership and geographical coverage.

Age distribution
Lifestyle databases tend to be under-represented among both the young and the old; and over-represented among intermediate age groups. This pattern is reasonably consistent between the sexes, although older men appear to be more responsive to lifestyle questionnaires than older women.

Car ownership
Response rates appear to be highest among middle ranking households. There may be a certain amount of scepticism about the benefits of completing the questionnaires among more affluent respondents.

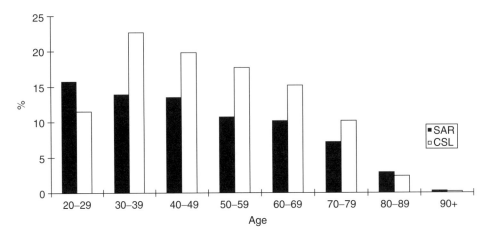

Figure 10.5 Age distribution of lifestyles data

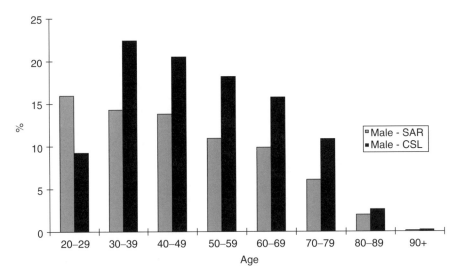

Figure 10.6 Car ownership within lifestyles data

Geographical coverage
Looks relatively balanced, although, perhaps, some holes in Northern England and the Borders. Likely to be more variation at an intra-urban scale, following demographic variations (i.e. tendency to under-representation in city centres, with concentrations among the young and poor, affluent suburbs and retirement areas).

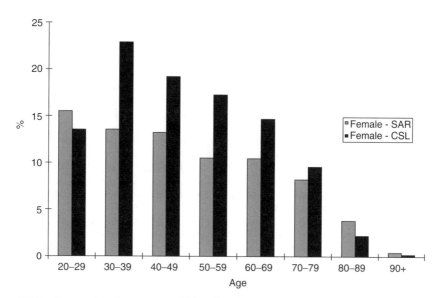

Figure 10.7 Geographical coverage of lifestyles data

If lifestyles data are significantly biased, then one possible strategy is to reweight individual responses according to their frequency in the database. To take a simple example, suppose a lifestyle database covers 50% of the population aged under 45, and 25% of the population aged 45 and over, then, in order to recreate the population from a lifestyles sample, we could reweight or duplicate each customer record twice for customers aged under 45, and four times for customers aged 45 and over. This approach was adopted by GMAP in the early 1990s to develop a geodemographic version of NDL's Lifestyle Selector (Birkin, 1993a, 1993b). The approach was later refined by Williamson, Birkin and Rees (1998) in an approach to reconstructing small area populations from Samples of Anonymized Records (see Section 10.5.1 below).

10.5 Other Micro-Modelling Strategies

10.5.1 Micro-simulation

Micro-simulation is a modelling approach that relies on the manipulation of individual data. It is, therefore, appropriate to lifestyles data or SARs, but cannot be used directly with spatially aggregate data such as census small area statistics. This approach is especially useful for forecasting, for database linkage and for problems that involve well-defined rules.

Database linkage

Williamson *et al.* (1998) developed a methodology for recreating 1991 census data for small areas from the SAR[1]. They used a genetic algorithm (GA) to extract samples from the SAR to match the characteristics of any small area. For example, extracts biased towards the unemployed or lone parents might be prevalent in poor inner city areas, whereas heavy selections from the professional classes or high car owning households might proliferate in the affluent suburbs. This approach aims to combine the strengths of the two classifications – 100% coverage from the census, and complete linkage between attributes from the SARs.

This work has commercial potential because it provides a method for linking or 'cross-matching' between any two databases, for example, a customer database could be linked to a lifestyle database to find the likely habits and preferences of customers. This might be used for new product development (what other products and services are the existing customers most likely to be interested in?) and promotion (what is the profile of customers most interested in the new product? Where are they?).

Rule-based approaches have been widely used in the estimation of benefits and impact of taxation (e.g. Orcutt *et al.*, 1986). When population characteristics are known, then well-defined sets of rules may be applied to determine outcomes, for example,

Characteristic	Outcome
Male aged over 65, not working	Assign pension
Child aged under 18	Pay child benefit
Income over £23 000 per year	Apply 40% marginal rate of taxation

and so on. Some commercial applications of this technology have been demonstrated. For example, certain banks have tried to monitor large monthly debits such as hire purchase repayments for motor vehicles. The most popular repayment cycle is 36 months. Therefore, customers can be targetted with an offer of finance shortly before their existing payment cycle is likely to end.

Projection and forecasting

These activities may be supported by either 'static' or 'dynamic ageing'. Once a micro-database has been produced, then the database may be aggregated and reweighted in line with some aggregate forecast, for example, growth in the elderly population. Having been reweighted, the data may then be disaggregated to the individual level once more. This is the process of 'static ageing'. On the other hand, demographic rules and probabilities may be applied. Thus, if an individual is 40 years old, that individual will be 50 years old in ten years time (rule), unless he/she dies in the meantime (apply probability of mortality). This is a process of 'dynamic ageing'.

Micro-simulation methods have been widely applied to public policy problems, such as welfare, hospital waiting lists and even water metering (see Clarke, 1996, for a review).

---

[1] The Sample of Anonymized Records is an extract from the census, in which all of the detail from individual or household records is preserved, but in which the zone of residence has been removed to protect the confidentiality of the respondent.

A micro-simulation approach to distribution channel modelling will be discussed in more detail in Section 10.6.

10.5.2 Psychographics

It is sometimes argued that the purchasing habits of customers are related neither to their demographic characteristics, nor to their lifestyles and interests. Examples such as the sales of designer clothing, organic foods and even fast cars are often quoted. It may be argued that certain purchasing decisions of this type may be dictated by deep-seated ethical views, or the desire for membership of some perceived group. The attempt to understand and manage this kind of behaviour is referred to as *psychographics*, and has particular relevance to media and advertising.

A notable example of 'psychographics' is the 'SocioStyles' system devised in France in the 1970s (Cathelat, 1993). In common with Lifestyles, SocioStyles starts from a questionnaire that asks individuals about their behaviour and attitudes but from a more emotional and cultural perspective. For example, one of the questions shows fourteen people dressed differently and asks the respondent: 'In an ideal world… which type of clothing would you like to wear?' (Cathelat, 1993, page 128).

The SocioStyles questionnaire was distributed more broadly, but less intensively than the major lifestyle questionnaires, with 24 000 responses from 14 European countries. The responses were analyzed and used to develop a classification, using statistical procedures that are very similar to the cluster analysis used to develop geodemographic systems. The visual presentation of the sociostyles classification is achieved using 'maps', which focus on two major factors that are labelled as 'values' and 'mobility'. Sixteen different psychographic types are identified (see Figure 10.8). For example, at what Cathelat refers to as the 'south-east' of this map, we can find individuals labelled as *strict*, *citizen* and *gentry*. As the names and their positions on the map imply, these are settled individuals

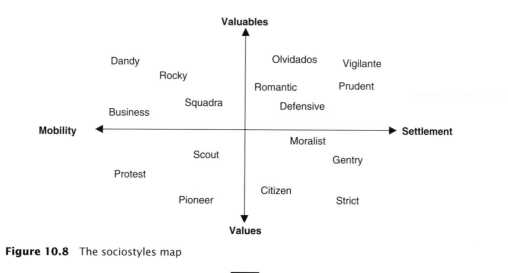

Figure 10.8 The sociostyles map

223

with strong moral values. At the other extreme, in the north-west corner of the map, we find mobile individuals who view economic wealth more highly than moral values. These people may be labelled 'dandy', 'business' or 'rocky'. Each individual type can also be described in something such as the sociostyles equivalent of a geodemographic pen portrait, for example:

Dandy – a hedonist youth with modest income seeking welfare structures
Strict – overtly repressed puritans in favour of social control

Sociostyles may be applied using panel surveys in which the purchasing patterns of individuals are tracked and linked to their psychographic characteristics. For example, Figure 10.9 shows a contrast between the users of washing powder and liquid detergent for a particular brand. Additional applications are seen as:

● Defining target populations for products
● Improved targeting techniques
● Identification of the best methods for production promotion (including use of media)
● Providing market information to promote product development and sales penetration

10.5.3 Customer segmentation

Many organizations now hold massive databases about their own customers, and these may be used to develop segmentations that are specific to that organization.

For example, WH Smith has a customer loyalty scheme that allows the company to track the purchasing behaviour of its customers, and to link this behaviour to the demographic characteristics of the customer (Quadstone, 2000). In common with lifestyle questionnaires, customers are also invited to specify their interests when registering for the scheme. Smiths have produced a segmentation on the basis of a combination of customer value and loyalty, purchasing behaviour (e.g. product preferences), customer interests and demographics. On this basis, customers have been segmented and profiled into nine groups: boy racers, mature browsers, bookworms, mother hens, back to school, couch potatoes, window shoppers, young spenders and magazine buyers. Each segment tends to have a distinctive profile in rather the same way that different geodemographic clusters have different demographic profiles.

In order to leverage maximum value from the loyalty card scheme, Smiths have conceived a framework for the recognition of customer value and loyalty:

Disciples	–	have a high transaction value and a high transaction frequency
Peripherals	–	have a low transaction value and a low transaction frequency
Browsers	–	have a high transaction frequency and a low transaction value
Core customers	–	have moderate transaction value and moderate transaction frequencies
Occasionals	–	have a high transaction value and low transaction frequency

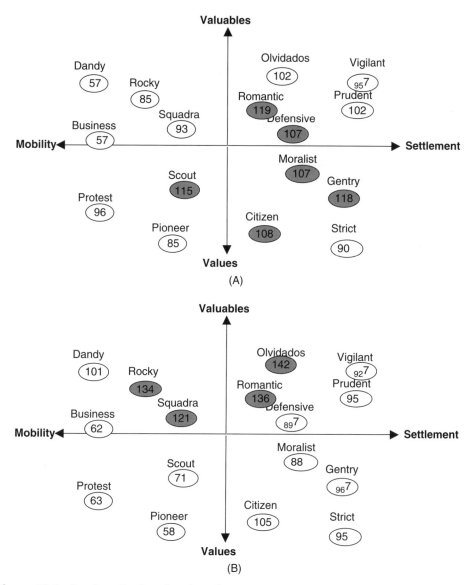

Figure 10.9 Panel application of socio-styles

Segmentation framework (schematic)

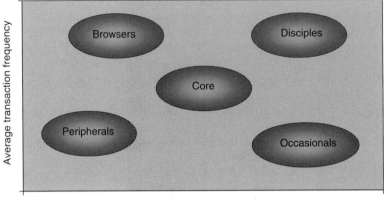

Figure 10.10 WHS customer segmentation

This framework may, in turn, be mapped onto the customer segments (Figure 10.10):

● Bookworms tend to be disciples
● Couch potatoes are often peripherals
● Magazine buyers, window shoppers and mature browsers tend to be either 'browsing' or 'peripheral'
● Boy racers and young spenders may well be occasional
● Back to school and mother hens are most likely to be core customers

The framework may also be mapped onto a plan of strategic customer actions (Figure 10.11). For example, disciples may need to be rewarded; cost reduction may be necessary for couch potatoes (e.g. withdraw loyalty cards or reduce floorspace in towns with high exposure to these customers). Browsers, especially magazine buyers, perhaps, need to be cross-sold a broader range of products.

It has been claimed that in order to implement a discounting strategy to build market share, WHS would need to reduce its profit margins by £56 million over five years; on the other hand, a micro-marketing strategy focusing on individual customer segments can add £48 million to margins over the same period.

10.5.4 Data mining

This approach has been much hyped in recent years. Data mining aims to use artificial intelligence (AI) techniques, such as neural networks, to find patterns in data (in preference to the kinds of clustering, techniques discussed in the preceding text). Two working definitions of data mining are as follows:

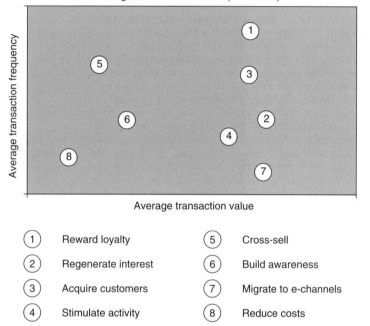

Segmentation framework (schematic)

Figure 10.11 WHS strategic customer actions

... the process of discovering valid, previously unknown and ultimately comprehensible information from large stores of data. You can use the extracted information to form a prediction or classification model, and to identify similarities between database records. The resulting information can help you to make more informed decisions (SPSS).

... referring to the use of a variety of techniques to identify 'nuggets' of information or decision-making knowledge in bodies of data, and extracting these in such a way that they can be put to use in areas such as decision-support, prediction, forecasting and estimation. The data is often voluminous, but of low value in its raw form, and little use can be made of it directly; it is the hidden information in the data that is valuable (IBM).

In practice, data mining can involve the combination of a whole range of different techniques from statistics, artificial intelligence or mathematical modelling. One of the key considerations is that many organizations have invested heavily in the concept of the data warehouse, in order to pull together information about the business and its customers into a single place. For example, in the early 1990s, banks and building societies held excellent databases about their current account holders, and could, therefore, identify characteristics such as average account balances, transaction frequency, and disaggregate all this according to the customer postcode or account-holding branch. They could do similar things for savings account or mortgage customers. Unfortunately, what was usually impossible was to connect any two of these databases, that is, to match individuals with a savings account to those individuals with a current account. Since the main objective of database marketing was to cross-sell new products to existing account holders, the inability to identify multiple product ownership was a significant impediment. Thus, data warehouses were built (at great expense!) to pull all of this information together. Retailers have been through a similar process, largely driven by the ability to capture huge amounts of customer information through the introduction of loyalty cards, and also by the ability of computerized stock control systems to track the sales of individual product lines, over time, in each individual outlet within a store network.

Now that the data warehouse has been established, it is natural that organizations wish to interpret and add value to this data. (It is also the only means by which the consultants, who have been well paid to introduce the data warehouse, can continue to justify their handsome fees!). Data mining provides the means by which this can be achieved. The interface to data mining software is typically its most impressive component, and three-dimensional data visualization is quite a common feature in the latest packages (see *www.quadstone.com* for a nice example). High performance computing technologies, including parallel processing, allow large volumes of data to be 'mined' relatively inexpensively.

One of the best-known packages is Clementine, a 'data mining workbench' within the SPSS product suite (SPSS, 2002). For example, GUS Home Shopping is the catalogue division of Great Universal Stores. (Home shopping businesses are among the most enthusiastic users of conventional geodemographic products, for example, for customer targeting and for assessing customer credit-worthiness.) GUS Home Shopping has previously used multiple regression models to forecast the demand for products in its catalogue in a season. What usually happens is that a 'preview' catalogue is mailed to a sample of customers before a new season begins. Customers are incentivized to buy from the catalogue. Demand forecasts for the whole customer population are constructed according to the characteristics and purchasing behaviour of the sample population. It is important to get these forecasts right, because if too much stock is held, then the excess will need to be discounted or written off; if not enough stock is held, then customers may be encouraged to purchase elsewhere. According to SPSS, GUS home shopping allocates budgets in the order of £200 million according to the demand forecasts.

Multiple regression estimates were often judged to be too conservative, leading to shortages of popular items and unhappy customers. Neural networks models were introduced to create 'forecasting models for around 6500 items using historical data from the previous three completed seasons.' The neural network model was tested directly against the existing multiple regression model using the same databases.

These experiments resulted in significant improvements to the forecasting process from using Clementine. The mean absolute error in the demand forecasts was reduced by 'up to forecast', and it is estimated that 'this capability produced a projected saving of 3.8 percent when compared with the previous regression method.' The increased accuracy of the forecasts also allows stock levels to be aligned more closely with customer requirements. ' By producing more accurate forecasts... service has improved and overstocking decreased.'

It is understandable that SPSS are keen to promote the benefits of their Clementine product. Others are less sanguine, and independent analysis of the commercial viability of data mining techniques is hard to find and evaluate. See and Openshaw (2001) has undertaken analysis of a response data set provided by a major bank to evaluate a wide variety of data mining approaches. He found that a complex neural network model was able to marginally outperform a 'linear model' (in effect, a gains chart as described in Section 10.2.2), but was also highly inefficient and computationally expensive. The case for data mining as a viable, let alone superior, alternative response modelling technique appears as yet unproven.

10.6   A Financial Services Channel Simulation Model

10.6.1   Introduction

Recent times have been characterized by an enormous proliferation in distribution channels (see Chapter 4). In other words, there are many alternatives means by which a product can be delivered from the manufacturer to the consumer, other than through the conventional route of a shop or bank branch. Thus, in retailing, we have seen growth in direct purchasing, especially through e-businesses such as Amazon, which may be seen as an essentially 21st Century incarnation of the mail order business; and for supermarket retailers, increasing use of home shopping and delivery services, again facilitated through the internet. In the case of financial services, in which there is no requirement for the distribution of a bulky physical product, a far greater number of options are available. GMAP (2000) have identified fifteen different channels to market financial services products (see Chapters 4 and 5). Other activities such as fuel retailing remain driven by conventional channels (i.e. petrol stations) for obvious reasons.

The growth of new distribution channels presents new challenges for modelling from a number of perspectives:

1. The introduction of new channels, in addition to the traditional routes, provides an extra dimension of complexity to the planning process – the question of 'how?' (are products and services accessed) in addition to 'who?' (are my customers), 'where?' (is my business successful), and 'what?' (are the customers buying) (cf Beaumont, 1991).

2. The introduction of new delivery methods introduces extra diversity to customer behaviour. For example, young customers are known to be the largest users of home delivery services, whereas older customers tend to remain wedded to traditional methods.

Both of these factors tend to work in favour of a *micro-simulation* approach to delivery channel problems. Micro-simulation is a technique that focuses on the characteristics and behaviour of individuals, rather than the groups that are used by conventional spatial interaction models (SIM) (see Chapter 10, Section 10.5). It is well known that micro-simulation provides an efficient means of representing complex problems (e.g. Birkin and M. Clarke, 1986), which, therefore, makes it effective for complex modelling. The micro-level representation also makes it suitable for modelling differences in channel behaviour among different socio-demographic groups. In the next section, we will look at the application of micro-simulation to the problem of channel simulation for retail financial services. We will present illustrations of the use of this model for both benchmarking and forecasting.

10.6.2 EC-Sim: A channel simulation model for financial services

The requirements of the channel simulation model are:

Step One: Build a micro-population that shares the demographic characteristics of the UK population.

Step Two: Add ownership of financial services products to this population.
In principle, these two steps may be thought of as a 'merging' process between the UK census (for the demographic data) and a market research database, such as the Financial Research Survey (FRS).

Step Three: Generate behaviour preferences for channel usage activities.
A simplistic view would be to see this as a further merging step to introduce data from the GMAP channel usage survey into the process. However, we need to be more sophisticated than this, because the behaviour that we can see in the channel usage survey is actually related, not just to demographics and product ownership but also to accessibility to services, which varies from place-to-place.

Step Four: Simulate channel usage behaviour on the basis of location, brand and demographics.

At this stage, we need to include measures relating to the physical provision of financial services, such as branch and ATM locations, and measures of accessibility between customers and services. Channel usage behaviour is modelled by small area at this step in the process.

In summary, the procedure, therefore, involves the integration of four types of data, relating to demographics, product ownership, channel usage and financial services infrastructure. These data are synthesized into a population list using a micro-simulation model, which also includes a spatial modelling component that adds a predictive modelling component to the channel usage process. This structure is illustrated in Figure 10.12.

We now examine the four steps in more detail.

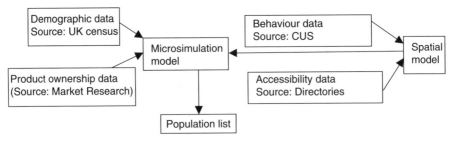

Figure 10.12 Overview of EC-SIM

1. Construction of synthetic populations: The objective of this stage in the process is to build a very large sample of individuals with characteristics that match the profiles of the UK population. Four questions that need to be addressed are:

How large to make the sample?
What attributes should be included?
How is the sampling to be performed?
Are there any other ways of doing this?

Size of sample

For the purpose of this chapter, we established a sample of one million households. This represents about 4.5% of the UK population. One of the advantages of current computing technologies is that it is now possible to construct and manipulate extremely large databases relatively easily. Indeed, one might easily conceive a list of 20 million households to represent the whole UK population. However, it needs to be borne in mind that we will wish to add a great many attributes to each individual household once the channel usage component is incorporated. This is likely to add hundreds of characteristics to each population and household record. A sample of one million households is a reasonable compromise between the desire to maintain a large sample, the need to manipulate this sample with reasonable ease and the need to add significant numbers of activity variables later on.

Attributes

The key demographic drivers of financial services product ownership are age, sex and social class. Secondary drivers of channel usage behaviour, such as educational attainment or car ownership, are not included as demographic attributes at this stage.

Sampling method

The sampling is conducted using a random sampling procedure known as 'Monte Carlo simulation'. The first step is to calculate a simple household quota for each UK postal sector. This quota represents 4.5% of all households in the sector. Then, we need to add age and sex of the household head, using known totals for each postal sector. These distributions are converted into the form of a contingency table that shows the probability that a household head falls within a particular cell of this matrix. This is, in turn, converted into a cumulative distribution of probabilities between, say, 0 and 1000. A series of numbers between 0 and 1000 are then pulled from a random number distribution.[2] The random number is matched to its position in the cumulative probability distribution. In this way, the age and sex of household heads is assigned at random, in accordance with the parent distribution.

---

[2] This part of the process is rather like spinning an enormous roulette wheel with 1,000 numbers rather than the usual 36. Hence, the derivation of the term *Monte Carlo* simulation.

The social class and exact age of the head of household are added in a similar way.[3] Then we look at the number of adults aged sixteen and over within each household, and once these adults are identified, then their sex, age and social class characteristics are also added. Children less than the age of sixteen are not included in the sample.

Alternative approaches

There are three other important ways in which a list of population attributes might be derived. The first is to use the SAR that provides a 1% population sample from the census. This, obviously, saves the trouble of synthetic sampling. Unfortunately, these could not be used in the work described here, because of issues in commercial licensing of the SARs. It should also be noted that the SAR is spatially aggregated to the level of metropolitan areas or local authority districts. A certain amount of estimation is still required, therefore, in order to obtain small area distributions. Applications of the SAR within micro-simulation modelling frameworks are considered in more depth by Williamson *et al.* (1998). A second approach would be to use a reweighted lifestyle database. This approach has not yet been attempted, to the best of our knowledge. Finally, it would be possible to use a financial service provider's own customer list as a basis for modelling. The principal disadvantage of this method would be that such a list represents a hugely biased sample of the universe of potential customers. The complications in trying to untangle these biases would probably outweigh any benefits from this approach.

2. Add ownership of savings products: Databases such as FRS provide information relating to the distribution of product ownership by age and social class. This data can be obtained in a spatially disaggregate form down to the level of UK postal areas. Since these market research surveys usually rely on a sample of a few thousand individuals, there is no way that further spatial refinement can be added with confidence. Thus, for adults within each postal sector, product ownership data is derived from ownership patterns within the postal area of residence.

Data might be obtained for a whole variety of financial services products, from current accounts to motor insurance or share ownership. Indeed, behavioural data about these channels and many others (for completeness: credit cards, mortgages, household insurance, personal loans, PEPs & ISAs, Tessas) are included within the channel usage survey (GMAP, 2000). For the purposes of this study, however, attention has been restricted purely to savings accounts.

For each individual within each household, we estimate the number of savings accounts held and the provider of each product. The market for savings account, at present, is characterized by multiple product ownership (i.e. individuals are quite likely to own two or more products) and also by fragmented branding, so that any one bank or building society may be offering a plethora of savings products with different rules on access, withdrawal, length of ownership, interest rates and so on. Our analysis concentrates on single providers rather than each of their products.

3. Customer behaviour: At this stage, we introduce behaviour from the Channel Usage Survey. Within the survey, no fewer than fifteen different channels are identified for

---

[3] For example, to add social class, we would look at another contingency table of social class by age and sex for each postal sector. The sampling process then proceeds as before.

financial services products, as we noted earlier. For the purpose of the channel simulation model, we have confined our attention to four major activities – branches, ATMs, telephone and internet.[4] However, all combinations of these channels are allowed; thus, while some users may still undertake all of their transactions through a branch, others may prefer to use the full variety of branch, ATM, telephone and internet. This gives a total of fifteen behaviour types, as shown in Figure 10.13.

Before commenting on the distribution of behaviour types shown in Figure 10.13, it is necessary to comment on one further complication within the channel usage process that relates to activities. Fourteen different transaction activities were identified by GMAP, including account openings and closures, deposits, cash withdrawals, withdrawals by cheque, transfer of funds, balance and statement enquiries, requests for information and requests for new product literature. All of these activities can be undertaken through a

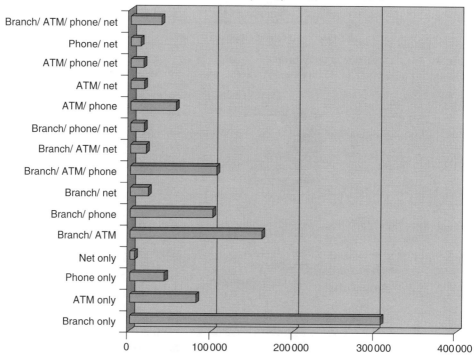

Figure 10.13 Channel usage behaviour types

---

[4] Note that each of these should be construed under the broadest possible definition. In particular, 'internet' should be thought of as a bundle of latest generation technologies, including WAP phones, interactive TV, palm top computers and so forth.

	Branch	ATM	Phone	Post	Internet	Int.TV
Withdrawal	21%	78%	0%	1%	0%	0%
Statement	17%	67%	6%	2%	9%	0%
Deposits	74%	19%	1%	5%	0%	0%
Transfers	56%	4%	18%	7%	12%	2%

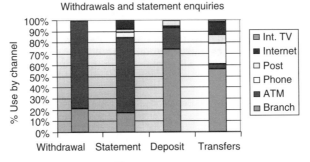

Figure 10.14 Transaction activity profile

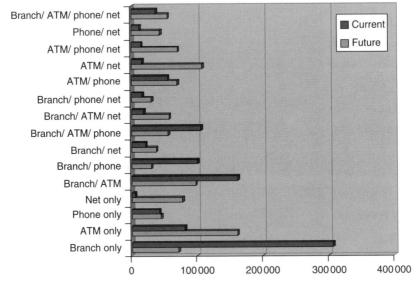

Figure 10.15 A behavioural scenario for channel management

branch. Most activities can be undertaken via telephone or internet, with varying degrees of ease. Balance enquiries and cash withdrawals can be undertaken most easily through an ATM. However, ATMs are not appropriate for activities that require personal contact or the physical exchange of documents, such as deposits by cheque.

ATMs have now overtaken branches as the dominant source of transactions. This reflects the importance of cash withdrawals, as the activity that accounts for 44% of all transaction activities (see Figure 10.14). However, 'branch only' remains the dominant behaviour profile in Figure 10.13, because as we have seen above, ATMs are only suitable for a limited range of transactions. 'ATM only' is, therefore, only possible for customers who only require a restricted range of services from their account provider. This analysis

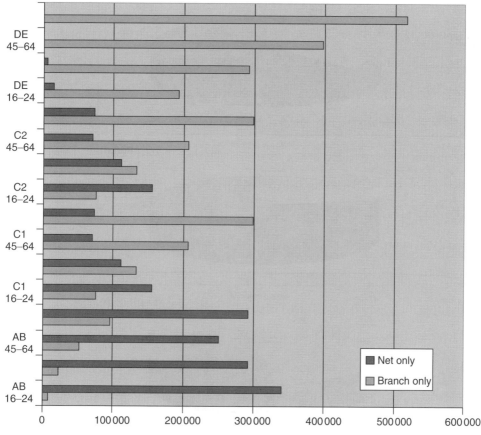

Figure 10.16 Channel preference by demographic group

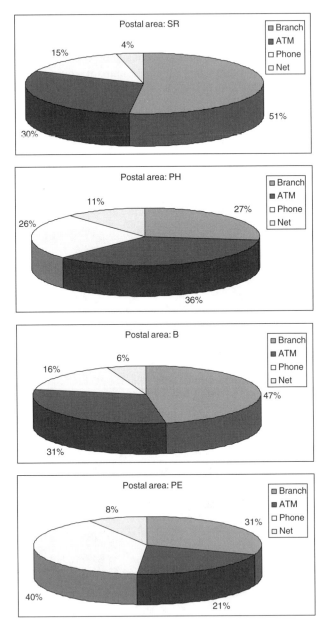

Figure 10.17 Channel usage by postal area

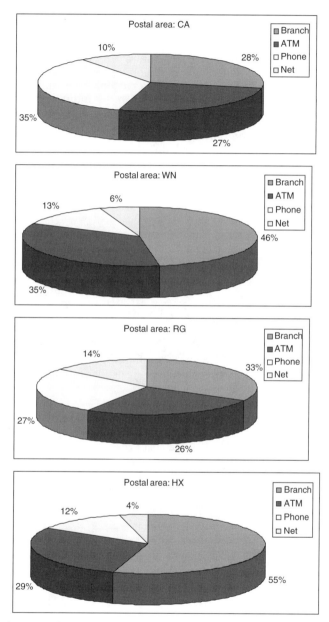

Figure 10.17 (*continued*)

also implies that ATM will be the dominant partner in behavioural combinations such as ATM branch.

In addition to existing behaviour profiles, we have also looked at how channel usage behaviour may be expected to change in the future. To an extent, this kind of activity always involves an element of guesswork, although it was also informed in this case by the inclusion of questions in the survey regarding customers' attitudes and expectations of the future. We can expect branch use to decline in the future, especially in favour of internet and ATM. This will be driven by a combination of:

- greater penetration of the appropriate technologies (e.g. internet, Wireless Application Protocol (WAP) phones) among the population
- by demographic processes that introduce more technologically aware customers into the adult population (it may be a cliche, but, nevertheless, substantially true that today's twelve year olds are more computer literate than their parents, and certainly than their grandparents)

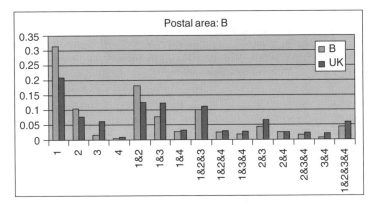

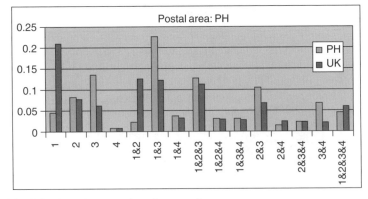

Figure 10.18 Behavioural transactions by postal area

● developments in the technology that allow more financial services activities to be undertaken by the newer technologies (for example, if an ATM could print cheques – securely – then there would be no need to go into a branch to have one written).

A future behavioural scenario including these trends is shown in Figure 10.15. This behavioural scenario underpins some of the forecasting applications, which are discussed later.

It is also important to emphasize that each of the behavioural patterns has its own socio-demographic underpinnings. Thus, the prevalence of different behaviour types will vary markedly between different demographic groups. This concept is illustrated in Figure 10.16, which shows demographic variations in expected usage between branch and internet for our planning scenario. Although stylized, the message that new channels

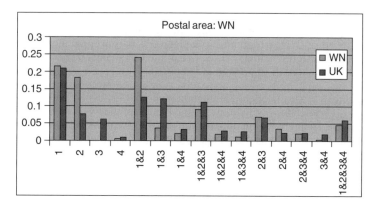

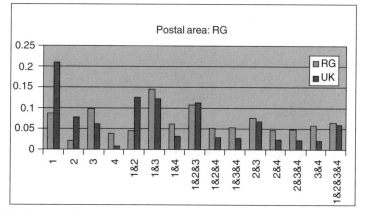

Figure 10.18 (*continued*)

would tend to be favoured by the young and affluent, whereas traditional channels would continue to be preferred by the elderly and less wealthy is a substantially accurate picture.

4.Transaction behaviour by small area: At this stage, the availability of distribution channels by small geographic area is considered. For branches and ATMs, in particular, the use of channels depends on their availability. Halifax customers in the north of England will tend to use branches and ATMs relatively heavily because Halifax has a high network density in this part of the country. In contrast, Halifax customers in the south and west will tend to favour channels such as telephone, post and internet because branches and ATMs are more thinly spread. These patterns would be reversed for an organization such as Woolwich Building Society, which is strong in the south and weak in the north.

The same effects will operate at different geographical scales. For example, if there is a Yorkshire Building Society branch in Yeadon, and no branch in Guiseley, then customers in Yeadon will be more likely to transact through branches than similar customers in Guiseley. In short, therefore, customer behaviour will vary, not only in relation to the social and demographic characteristics of the customers but also in relation to the configuration of financial service provision. By a similar logic, changes in provision can be assessed. Thus, if a branch is closed in 'Mytown', then customers will migrate, partly to other branches, partly to other distribution channels. Thus, the impact of network changes on transaction behaviour can be evaluated at this stage of the process.

10.6.3 Planning applications

In this section, we begin with an illustration of the baseline outputs of the Channel Simulation Model. These kinds of analysis could be used by financial services providers to

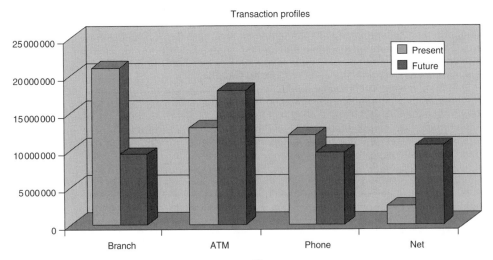

Figure 10.19 MyBank scenario transaction profiles

plan better delivery channels, for example, to understand where the mix between branches and ATMs within an area is right, and where the balance is wrong (see also Chapter 11), to set better targets for their branches that allow for local market effects, or to educate customers in an effort to migrate them between high- and low-cost channels.

Figure 10.17 shows illustrations comparing channel usage patterns for each of the four channels in two contrasting postal areas. The purpose of these illustrations is to hammer home the point that the combination of channel availability, customer demographics, and regional infrastructure gives rise to significant variations between places for all channels. In

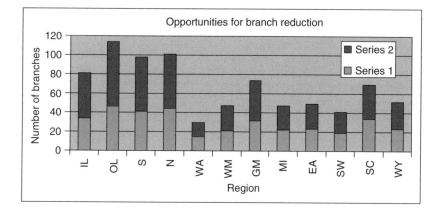

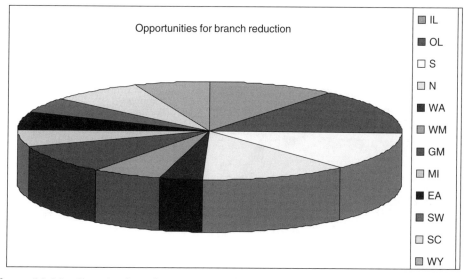

Figure 10.20 Changing branch requirements by region

the case of branch usage, Sunderland (SR) is compared with Perth (PH). Twofold variation in branch usage can be seen between the two areas, with much of the difference accounted for by the adoption of the newer technologies (phone and internet). Note that this is, probably, a combination of push and pull factors – customers in Perth may be slightly younger and more affluent than their counterparts in Sunderland; but equally, the residents of Perth may be forced onto newer technologies by the difficulty in accessing a sparse network of bank branches. A third interpretation might rest with market share variations: if the banks that dominate in Perth (e.g. Bank of Scotland, Clydesdale, Royal Bank of Scotland) provide

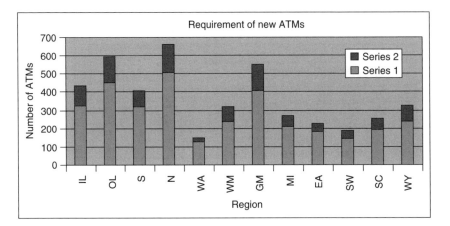

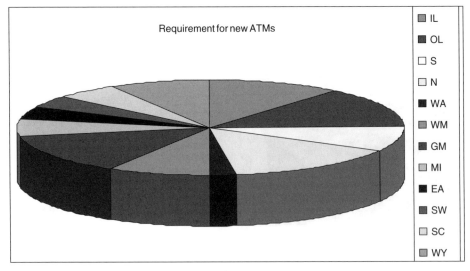

Figure 10.21   Changing ATM requirements by region

better new technology services than their southern counterparts. Similar illustrations are provided for ATM usage, between Birmingham (B) and Peterborough (PE), for telephones between Carlisle (CA) and Wigan (WN), where there is a fully threefold variation, and for internet uptake between Reading (RG) and Halifax (HX), where the proportional variations are even greater.

The combination of behaviour profiles for four of these areas are shown in Figure 10.18. For example, we can see the heavy preponderance of branch-only transactions in Birmingham, whereas Wigan is notable for the combined use of branches and ATMs. The phone is well established in Perth, both alone and in combination with branch, ATM or internet. Sophisticated customers in Reading tend to be averse to branch and ATM, and in favour of telephone and internet.

The remainder of the illustrations in this section refer to forecasting applications for 'MyBank' as a provider with strength in the north of England, which might equally well be Yorkshire Bank, Yorkshire Building Society, Halifax plc, Bradford & Bingley or Northern Rock. No specific policy conclusions should be drawn in relation to any one of these providers, or any other. Figure 10.19 shows how the transaction profiles of MyBank customers will change under the planning scenario described in Section 10.3.3, previously. The changing balance of requirements between branch and ATM for this scenario is illustrated in Figures 10.20 and 10.21. This example uses Standard Regions as a base for clarity in representation, because there are sufficiently few of these to incorporate within a single illustration. Figure 10.20 shows that large numbers of branches need to come out right across the network, but the opportunities for branch reduction are greatest in areas such as London and East Anglia, where more than 60 branches in every 100 would become redundant if our planning forecast becomes a reality.[5] However, these branches do need to be replaced by greater numbers of ATMs, with the greatest requirement falling in London, Greater Manchester and the North. This is partly because the growth of internet as a substitute distribution channel is expected to grow less fast in areas such as London and Greater Manchester than in places such as Wales, East Anglia and the South East. These trends are summarized in Plate 6.

10.7 Conclusions

In this chapter, we have seen that there are many ways in which the process of direct selling to individual customers may be supported. Many of these methods rely on an understanding of the unique characteristics of individual customers. These characteristics may sometimes be modelled through a process of *micro-simulation*. We have argued that the technique of micro-simulation is particularly appropriate for the management of distribution channel networks.

---

[5] It is worth emphasizing that we are not necessarily advocating wholesale branch closures within the financial services industry. These are simply illustrations of changes, which could be driven by customer behaviour in the future. We should also note that just as model-based planning supports decision-making in business, it is also possible to inform the regulation of businesses in the same way (see also Chapter 11).

MEASURING THE BENEFITS OF SPATIAL MODELLING

11

11.1 Introduction

In the preceding chapters, we have demonstrated that geographical modelling techniques may be applied to a variety of problems in retail and service planning. These problems include territory planning, sales forecasting and impact analysis, network design and direct marketing. Along the way, we hope to have demonstrated that applied research of this type presents significant challenges, and that solving these challenges in itself can lead to significant refinement of existing planning techniques, and to the development of new methods.

While this line of argument is of interest to students and academics, the major driver behind these developments is commercial as much as intellectual. In other words, we have been driven to find new and improved solutions to business problems because of the value that these solutions provide. In this chapter, we wish to focus explicitly on the nature of this value. The remainder of the chapter will be structured around a framework that considers four kinds of business benefits. In Section 11.2, the relationship between accurate sales forecasting and returns from capital investment will be explored. The exploitation of links between geographical intelligence and operational efficiency will then be discussed. Examples from the financial services sector will be used to demonstrate the capability to extract competitive advantage in Section 11.4. A fourth class of benefits relating to strategic planning will be considered in Section 11.5. Having focused heavily on the commercial benefits of applied modelling, the last substantive section of the chapter will consider potential applications and benefits within a public planning context. Some other kinds of benefits, together with concluding comments, will be offered in Section 11.7.

11.2 Returns on Capital

An obvious starting point for this discussion is to consider the problem of locating a new retail outlet. As we saw in Chapters 7 and 8, it is possible to construct 'what if?' planning models that allow the impact of this type of development to be simulated with considerably greater accuracy than conventional techniques. This allows the likely turnover to be assessed with confidence. A particularly important benefit that can arise from the use of spatial interaction models (SIM) is that the cannibalization of trade from existing stores

Table 11.1   Key development performance ratios

	Base	Scenario	Uplift
Turnover	20	22	10%
Fixed costs	5	5	0%
Variable costs	12	13.2	10%
Profit	3	3.8	27%
Margin	15%	17%	15%

can also be assessed (see Chapter 8). The overall effect is that the net benefit of the decision to invest capital in building a new outlet can be quantified with increased precision. This, in turn, means that better sites can be identified, and better profits can be realized.

A simple example is presented in Table 11.1, in which we have supposed that the adoption of appropriate planning methods has allowed us to obtain a 10% increase in turnover by getting the store opening in the right place. In line with definitions adopted previously (see Chapter 8) it is natural to think about the outcome of conventional planning techniques as a 'baseline' model; and the result of applying advanced spatial modelling techniques as a 'scenario'. Thus, the baseline value for a new supermarket might be £20 million per year, with a scenario value of £22 million.

Let us further suppose that the maintenance of the new facility requires a fixed cost of £5 million per annum. These costs must be met regardless of the trading performance at the new site. On the other hand, variable costs are proportional to trading levels. In the example, we have assumed a variable cost of £60 for every £100 of turnover. The uplift in variable costs is, therefore, directly proportional to the uplift in turnover–as turnover increases by 10% from the baseline to the scenario, so the variable costs also increase by 10%. Profit can be derived as turnover less both fixed and variable costs. Since the difference in turnover between the two cases exceeds the difference in variable costs, and since the fixed cost is the same in both scenarios, the profits are uplifted by £800 000 in the scenario, an increase of 27%. This also represents an uplift of 15% in the gross operating margin. We can generalize this example to conclude that as long as:

1. there is some fixed cost in operating a retail site and
2. it is possible to find better sites using spatial modelling techniques than by existing techniques

then the application of spatial modelling will provide business benefits in the form of:

● increased turnover
● more profit
● better margins

Note that clause 1 will always hold in an applied context, not least because the cost of investing in a new outlet needs to be amortized over a period (compare the discussion of Chapter 8, Section 4), which in itself implies that there is a fixed cost to be met. Furthermore, the arguments of Chapters 7 and 8 tend to support the assertion that clause 2) will generally hold.

Let us look next at a slightly more complicated example. GMAP has developed site rating systems for financial services customers, using similar principles to the fuel ratings described in Chapter 7 (Section 7.4.3) (GMAP, 2001). The objective is to provide location scores for cash dispensing machines (ATMs), using a variety of inputs, such as local demographics, workplace populations, competing locations (other ATMs, bank branches, post offices and so on), and retail adjacencies (reflecting variations in customer 'footfall' between different locations according to the quality and density of retail activity in the vicinity). These location scores tend to be highly predictive of activity levels for each machine. A typical example demonstrating this correlation through straightforward regression analysis is shown in Figure 11.1. Both the location scores and the associated 'real' transaction volumes are, typically, normally distributed with high levels of performance variation, as illustrated schematically in Figure 11.2.

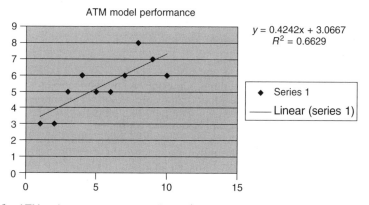

Figure 11.1 ATM ratings versus transaction volumes

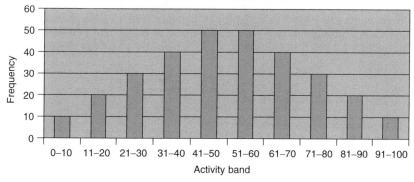

Figure 11.2 Distribution of ATM transactions

247

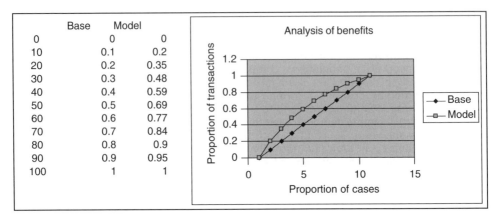

	Base	Model
0	0	0
10	0.1	0.2
20	0.2	0.35
30	0.3	0.48
40	0.4	0.59
50	0.5	0.69
60	0.6	0.77
70	0.7	0.84
80	0.8	0.9
90	0.9	0.95
100	1	1

Figure 11.3 A benefits curve for location selection

Suppose that the model is used to identify a large number of ATM locations from an even larger universe of potential locations. Although the model may not be perfect, it will be possible to identify better locations than from simple random selections. This principle is illustrated graphically in Figure 11.3. The ability to exploit the value from this modelling approach will vary according to relative positioning on the benefits curve, which may, in turn, be related to factors such as network maturity, market share and coverage. Thus, for a well-established brand the number of potential sites may be relatively restricted, which reduces the scope for selectivity between locations. However, for immature brands, the number of potential sites may be extremely large relatively to their requirements. A good example of an immature brand would be one of the many surcharging providers currently seeking to establish a foothold in the European financial services market (e.g. Green Machine, MoneyBox).

If the average activity level is 10 000 transactions per machine, and we need to service an additional one million transactions, then we would need 100 new ATMs. Looking at Figure 11.3, if 'model selectivity' is running at 50%, then we would only need 85 ATMs to service this business. If model selectivity can be improved to 20%, then we would only need 70 ATMs. Assuming that an ATM costs £40 000 to install and £20 000 per annum to maintain, this yields a net present saving of approximately £5 million at 20% selectivity.

11.3 Operational Efficiency

The ability to streamline operational processes through automation within large organizations was one of the earliest drivers for the adoption of Geographic Information Systems (GIS). For example, Mahoney (1989) argued that the introduction of digital maps and terriers (i.e. facility inventories) could reduce the time and money cost in accessing infrastructure data within local authorities by 85%: that is, they are able to spend £15 where £100 was required previously.

It is likely that the advent of mobile telecommunications technologies will yield still further benefits of this type. For example, consider once again the fuel ratings described in Chapter 7. We saw that these systems depend on the collection of detailed information about each individual petrol station within a network. The process might be characterized by the sequence of steps shown in Figure 11.4(A). First of all, a general form must be designed; then the forms are completed at each site. Each form must be processed to provide a network database. Next, an information system must be designed to absorb the data. Once this is done, the data may be imported for manipulation and display within the system. Only on completion of each of these preparatory steps is it possible to calculate and display ratings for each site.

Compare the situation assuming that the data for each site may be captured using a hand-held or 'palm-top' computing device. Now the exercise can be reduced to three simple steps–the design of a rating system, the creation of a data entry mechanism from the hand-held device, and data entry for each site (Figure 11.4B).

Most business applications of retail intelligence and network planning are less clearly process-focussed than these examples. Nevertheless, other benefits of an operational nature can also be identified. Looking back to Chapter 9, we saw that a combination

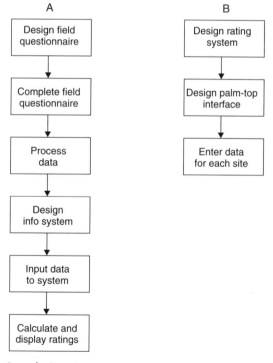

Figure 11.4 Fuel ratings design steps

of enhanced computational power, perhaps in combination with parallel hardware, has facilitated the solution of ever more complex network planning problems. Another example of this type is illustrated in Plate 7. Here, a spatial interaction model has been constructed to simulate expenditure flows between very small geographical areas (enumeration districts) and retail destinations. In a typical application, there might be 150 000 EDs and about 1000 retail destinations. If the model is further disaggregated by ten product lines, say, and six customer groups, then this means that a total of:

$$150\,000 \times 1000 \times 10 \times 6 = 9000\,\text{million}$$

individual flows must be calculated for each model run.[1]

What we do next is to simulate the effect of opening a new store in each of the enumeration districts in the country. This means that we run the model–with 9 billion flows–150 000 times, once for each ED. The results of this process are mapped as a surface in Plate 7, with the red zones showing areas of highest potential for new supermarket development.

The point is that the ability to process huge numbers of spatial analysis operations using contemporary computing resources can bring fundamentally new applications into range. In this case, a retailer can reverse the reactive process by which development opportunities are appraised on a case-by-case basis. Instead, it is possible to search through the universe for all possibilities, and then begin to search proactively for suitable opportunities.

11.4 Improved Competitiveness

Retail businesses are typically driven by some combination of two objectives:

to maximize market share, and ideally to dominate markets
to increase the profitability of the business

Both of these objectives can be promoted through the deployment of Spatial Decision Support Systems.

Figure 11.5 shows an example of the outputs from GMAP's Channel Usage Survey (GMAP, 2000–see also Chapter 5, Section 5.4). The example shows the contrast between two different regions of the country. The South-West is a rural area, sparsely populated with few major towns and an economic dependence on farming and tourism. In parts, the North-East shares these features, but has a much larger concentration of urban centres (Newcastle, Middlesborough, Sunderland), with a historical dependence on mineral extraction and manufacturing, and high levels of unemployment. These differences tend to be reflected in the distribution of financial services customers within the two regions. The customer distributions are illustrated in the three-dimensional pie charts.

---

[1] In practice, it is not quite as bad as this, because we can ignore a lot of the flows from one end of the country to the other. Nevertheless, the number of calculations to be performed still has several noughts on the end!

1. 'Already Migrating'. Customers in this segment have efficient usage patterns, under-taking at least 70% of their transactions via low-cost channels. For the purpose of this analysis, we have defined ATM and internet as low cost channels.
2. 'Branch dependent'. Customers in this segment will undertake at least 50% of their financial services transactions via the branch.
3. 'Confetti customers'. These customers have unusually high activity levels, averaging at least twenty financial services transactions in a three-month period.
4. 'Doing without'. Type D customers have extremely low activity rates and product ownership levels.
5. 'E-adopters'. This group of customers is at the vanguard of PC and internet take-up. All of the customers in this group have purchased some product or service via the internet.
6. 'Floaters'. These customers cannot be classified under any of the five definitions presented above.

In the North-East, we can see that the dominant customer types are A and B. Groups C and E are less important, with types D and F adopting an intermediate position. These patterns are reflected in the channel preferences of customers in the North East. The large number of type A customers is translated into a heavy utilization of ATMs for financial transactions. However, the equally large preponderance of type B customers feeds through into high levels of branch usage also. Other channels, particularly internet, are of little significance.

On the other hand, the dominant customer groups in the South West are Ds and Es. The large number of 'E-adopters' gives rise to a small but significant bundle of internet transactions. The large proportion of customers 'Doing Without' yields transaction volumes, which are substantially lower than in the North East. The average number of transactions per quarter in the South West is around 7, which is less than half of the average of 16 in the North East.

Ultimately, when the transaction profiles of the regions are compared (in the charts at the bottom of Figure 11.5), we can see two substantially different profiles. The South West is characterized by a dependence on ATMs, low overall activity levels, and the beginnings of significant e-business transactions. The North East is characterized by a balance of activity between branches and ATMs, high overall transaction levels, and little evidence for migration towards emerging distribution channels.

And yet, for all these differences, the comparative channel profile of the two regions is striking in its similarity, with a ratio of approximately five ATMs to every three branches. In fact, the provision of channels per million customers is also about ten percent higher in the South West. Clearly, these patterns do not adequately reflect the underlying customer activity levels, on which basis one would expect more branches in the North East, fewer branches in the South West, and lower overall provision levels in the South West. We are inclined to conclude that if financial services providers had access to better-quality market information, they could obtain competitive advantage through improved alignment between the provision of services and the needs of their customers. Furthermore, the impact of alternative network structures could be evaluated, and potentially optimized, through application of the methods described in Chapters 7–10.

A second example of competitive advantage concerns customer attrition, also commonly referred to as customer *churn*. Again, the data in this example relates to financial services. We are looking at the customers of a large UK bank ('MyBank') who held mortgages on 1st January, but were no longer mortgage customers on the following 31st December. The

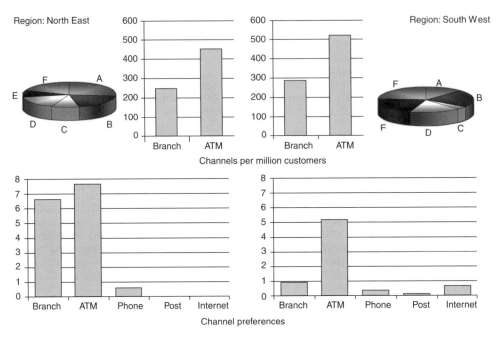

Figure 11.5 Comparative channel profiles for two UK regions

charts in Figure 11.6 show the profile of these 'disloyal' customers. We can see that each of these criteria displays a clear pattern.

Levels of customer churn are highest in towns with the lowest market share for the MyBank network, reflecting higher levels of customer loyalty in zones of high brand strength. Higher levels of churn are found among customers with large mortgages. In part, this may be a reflection of the fact that these customers have more to gain by switching to a lower cost provider. It may also be partially related to higher levels of sophistication among well-informed customers who have the means to sustain higher levels of mortgage expenditure.

Low accessibility yields high churn. This analysis is based on average travel time between customers and their account-holding branch. Even though mortgage transactions are rarely conducted through a branch (much more frequently by post or telephone), there is, nevertheless, a systematic relationship between accessibility and churn.

Branch reduction yields high churn. These patterns were established by comparing rates of attrition in centres that experienced changes in representation across the study period–typically the closure or merger of a local branch.

Customer attrition by social class shows similar patterns to attrition by mortgage value, and probably for the same kinds of reasons, that is, high social class customers are likely to be more highly educated and well informed: and to have invested more in their mortgages.

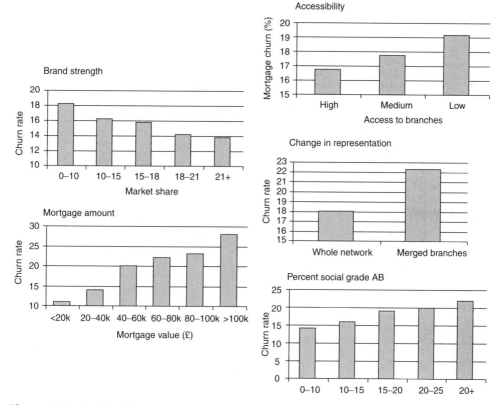

Figure 11.6 Profile of mortgage customer churn

These factors may be combined into an attrition model that is predictive of churn for either individual customers or small geographical areas. A notional example is presented in Table 11.2, in which two different types of area are contrasted. 'Coketown' is conceived as a place with a strong blue-collar-manufacturing base, which lies within the traditional regional heartland of the MyBank group. Mortgage borrowing is typically low against similarly modest household incomes. The dominant lifestage is the nuclear family. These characteristics imply low levels of customer attrition at about 8%. 'Pleasantville' is conceived as a town with a young, affluent and professional population. Households are heavily mortgaged on the basis of substantial incomes. However, the town is well away from the traditional base of MyBank, and is experiencing reduced service levels as a result of branch rationalization. These characteristics imply high levels of customer attrition at about 30%.

Through improved understanding and modelling of these patterns, it is likely that better strategies can be adopted for customer retention. It may also be possible to extend the analysis to target potential customer attrition towards MyBank and away from other

Table 11.2 Contrasting customer loyalty profiles

Attrition model Example		
Location	Coketown	Pleasantville
Provision	High and stable	Low and receding
Income	£15 000	£100 000
Mortgage	£10 000	£150 000
Occupation	Blue collar	Professional
Family	Wife and 2 children	None
Risk of churn	8%	30%

Copyright @ 2000 GMAP Ltd.

brands. Clearly, both of these activities imply competitive advantage. Specific tactics might include targeting customers at high risk of churn, with a view to retaining their business, or adding mortgage advisers or improved facilities in branches at locations such as Pleasantville to protect market share in these places.

11.5 Strategic Value

One of GMAP's earliest clients was WH Smith. At this time, 'GMAP' was very much a cottage industry, drawing on academic consultants in their 'spare time'. We began by dividing the United Kingdom into 31 planning regions, starting with a pilot study for a single region (we were allowed to choose Yorkshire!). Once our models had been extended and applied in a number of regions, GMAP was invited to make a presentation to the main Board of Directors of the company. Among many more sophisticated examples, we presented one piece of analysis that was particularly well received. This was simply a map of the United Kingdom showing Smith's penetration of key markets (principally books, stationery, 'sounds', and greeting cards). The map showed a consistent pattern of high market penetration in the south, declining steadily towards the north. Our clients were profoundly impressed by the opportunity for building turnover simply by reproducing observed levels of performance in certain parts of the country on a much wider geographical scale. This became a fundamental part of the company's development strategy for a number of years, culminating in the purchase of the rival Menzies chain, with a strong Scottish presence in the early 1990s.

Table 11.3 Network development potential for major super-market groups

	Market share		Uplift
	Baseline	**Scenario**	
Tesco	17.85	22.21	24.4%
JS	14.77	19.11	29.3%
Asda	12.15	15.48	27.4%
Safeway	8.49	10.72	26.3%

What goes for Smiths is equally true for the majority of retail and service sectors. Even the major players, typically, have highly irregular levels of market penetration and performance from a combination of historical brand weakness and uneven distribution networks. This point is reinforced in Table 11.3. Here, we have analyzed the market performance of key players in a number of retail and service industries. In the 'baseline', we have considered market share at the postal area level, that is, breaking the country down into 120 constituent areas. Then we have considered every postal area where performance yields a market share below the average, and what happens if this can be lifted to the existing average. In other words, performance in all of these areas is still no better than it is across the remaining half of the country. These results are shown in the 'scenario' columns of the table. Although these markets are competitive, and such levels of performance cannot be achieved by all the businesses simultaneously, they, nevertheless, demonstrate what is clearly achievable if the problem of network development is approached systematically.

Since the process of significant network development requires significant investments of time, money and management expertise, many companies have sought to take a shortcut to this process through the merger of complementary networks, or the acquisition of rival businesses. This mergers and acquisitions (M&A) process is another area in which strategic value may be extracted. An example, again from the financial services industry, is shown in Figure 11.7 (cf. Birkin, Clarke and Douglas 2003). Here we can see a situation in which two businesses ('A' and 'C') have been joined following a merger. This creates a situation in which there are four branches in the highlighted centre, two branches from the newly combined business and two competitors (labelled 'B' and 'D'). The illustration shows eight different reconfiguration strategies, which might be adopted in this context, including the retention or closure of existing branches, or from the introduction of new branches (labelled '3' in the diagram). For example, if branch 'C' is closed (scenario 2), then market share falls from 15% to 10%, and profits are also reduced by a third.

This kind of analysis is particularly important to banks and building societies that have tended to adopt a fixed mind set that two or more networks may be merged with the removal of overlapping branches, yet having no effect on the operational performance of the combined businesses. In fact, the options are even more complicated than shown in Figure 11.7, if we consider the universe of reconfiguration strategies in addition to simple openings, closures and mergers. In this context, reconfiguration might mean, for example, the complete or partial automation of a branch in which existing service points are replaced by ATM or other electronic facilities. In practice, any number of formats may be permissible for any particular outlet, and, hence, the problem of optimizing formats or

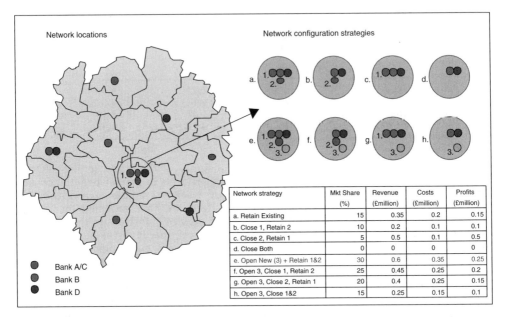

Figure 11.7 Financial centre network configuration strategies

configurations becomes quite complex even when applied to a relatively simple case, such as Figure 11.7. When these options are multiplied across whole networks, then the need for sophisticated combinatorial optimization tools, such as those illustrated in Chapter 9, starts to become apparent.

Network planning models can also be used at other stages in the M&A process, besides the integration of distribution channels post-acquisition. Specifically, the issues of target identification and financial evaluation are considered by Birkin *et al.* (2002).

11.6 Applications to Market Regulation

One other range of benefits, which are worthy of independent consideration, is the use of network planning models for the regulation of retail activities. Much of the thrust behind the original development of spatial interaction modelling approaches in the late 1960s and 1970s came from a desire to use models within an applied urban modelling context. This stimulus was evident in both the United States (see, e.g., Boyce *et al.*, 1970, for a review of a number of applications) and in the United Kingdom (an excellent review is provided by Batty, 1989).

Such approaches fell into disuse after the mid-1970s in the United Kingdom, in part, due to critiques of the effectiveness of the methods (e.g. Sayer, 1976), and, in part, due to changes in the political culture of the country following the election of the first

Thatcher administration in 1979. We have sought to demonstrate throughout this volume that models do now provide effective planning tools, and if this is true for commercial users, then it can be equally true for public sector planners. Within the governmental planning process, there is now renewed interest in the control of both major out-of-town developments, such as the Meadowhall, Merry Hill and Bluewater Park schemes approved in the 1990s; and within more local planning contexts, that is, whether individual towns need competition, say from a new supermarket, or whether existing traders should be protected from new development.

Two potential planning applications of the models discussed here might be to 'competition policy' and to 'local market development'.

1. In the United Kingdom, the Competition Commission (formerly the Monopolies and Mergers Commission) and in Europe, the equivalent regulatory body are both charged with maintenance of fair play within competitive markets. There is increasing realization that unregulated markets may not always lead to an ideal outcome from the customer's perspective. This is particularly true when mergers and acquisitions may serve as a precursor to network rationalization and service reduction. Excellent examples of this are provided by the wave of mergers in the UK financial services industry over the last five years, that is, Halifax with Leeds Permanent, Halifax with Birmingham Midshires, Barclays with Woolwich, Lloyds-TSB-C&G and latterly Royal Bank of Scotland-NatWest. Such mergers are significant because they do automatically trigger interest from the Competition Commission, which has the power to impose real restrictions on the organizations concerned. A good example would be the requirement for Granada to sell off a significant number of motorway service stations in the United Kingdom, following its acquisition of the Trust House Forte group in 1998. It can be seen that the kind of analysis discussed in Chapter 9, Sections 9.3, and 9.4, could provide a strong analytical foundation for interventions of this type. For example, if the merger of Barclays and Woolwich provides a significant reduction in choice for the residents of (say) Sheerness, then maybe the protagonists should be required to adopt some ameliorative strategy–perhaps to divest one of these branches, or to keep them both open.
2. In local markets, one of the problems is the lack of adequate benchmarks against which appropriate development criteria may be judged. For example, Langston et al. (1997), Poole et al. (2002a, 2002b) have demonstrated huge variations in retail market saturation across geographical areas. This could provide one criterion for the judgement of planning applications (i.e. to favour new applications in relatively unsaturated markets). Considering the types of indicators introduced within Chapter 9, Section 9.3, one could easily extend this argument, for example, to look at local market shares (e.g. a retailer might be restricted from bidding for new development opportunities in a town where its share already exceeds, say, 30%), or to encourage new development, where the existing access to opportunities is low.

Having demonstrated the value of spatial modelling within a commercial environment over many years, the ability to transfer the technology back into the kind of public planning environment from where it originated represents a major opportunity.

11.7 Conclusion

In the previous sections of this chapter, we have considered the potential benefits from applied spatial modelling under the four headings of capital investment, operational effectiveness, competitive advantage and strategic analysis. We have presented a number of examples that illustrate these benefits, from a variety of business sectors. It should be emphasized that each of these kinds of benefit will typically accrue from applications in a whole variety of different sectors including retailing, financial services, petrol retailing, automotive vehicles distribution and telecommunications. The use of exhaustive illustrations from each of these sectors, in the context of each benefit, is prohibited only by the combination of time and space.

Although many other commercial benefits beyond those illustrated within this chapter may be conceived, it is probable that the majority may be located somewhere within the framework that has been proposed above. In fairness, it should, perhaps, be noted that with so many frameworks, reality is actually a little more cloudy. Thus, many benefits may actually straddle more than one of our four categories, in contrast with the exclusivity implied in our previous discussion. For example, Birkin, Boden and Williams (2002) provide a discussion of the benefits of site assessment for capital allocation between countries. This might be characterized as a benefit in relation to returns on capital, or competitive advantage; although operational efficiency could be the most appropriate category of all. The ability of spatial models to recommend optimal formats at specific locations might be regarded as beneficial to returns on capital; although when the equivalent technology is applied across a network, the value has a more strategic flavour as we saw above (Section 11.5). The use of spatial models to target advertizing spend, following a new store opening or refurbishment, should probably be characterized as a contribution to operational efficiency. The use of models to support market entry strategies, for example, for a retailer planning to enter Poland or Hungary for the first time, might be considered to provide both competitive advantage and strategic value. Consideration of further examples of this type is left as an exercise for the reader.

CONCLUSION

In the book, we have focused on the development and commercial application of techniques for 'retail intelligence and network planning'. Many of these applications have come from our own research and consultancy work over the last fifteen years. It is difficult to overemphasize the revolutions in technology that have helped to drive forward, to facilitate and to shape this process. For example, an often-quoted statistic is that computer power has tended to double once every eighteen months over the last fifteen years. This means that things that were difficult to achieve on mainframe computers fifteen years ago are now relatively straightforward on a laptop. Good examples would be things such as microsimulation or optimization of a network with complex customer interactions. For example, in the late 1980s, two of the present authors developed a microsimulation approach to a population and income estimation problem using a synthetic sample of 50 000 households (Birkin and M. Clarke, 1988, 1989). In order to run models, our programs were run overnight in batch mode via competition for resources with many other university departments on a mainframe computer costing about £1.5 million (in 1980s prices!). In Chapter 10, we reported on an application to the financial services market, which uses a sample of one million households, and can be run in a few seconds real time on a personal computer costing around £1000.

Another example can serve to emphasize the same point in a different way. As recently as the late 1980s, we were producing maps of the type illustrated in Figure 12.1. The maps were processed using a 'gwbasic' compiler, without the use of so much as a digitizer tablet. Instead, the boundaries of each area were constructed from a series of directional commands, including 'u1' (move up one pixel on the screen), 'd2', 'l2', 'r2' (move down, to the left, or to the right two pixels), and similarly e3, f3, g3, h3 (comprising movements of three pixels along any of the four diagonals). Several months of student labour were invested in the creation of a suite of postal sector maps for regions of the UK. Despite the relative crudeness of these maps, they served their purpose as a basis for the representation of patterns. Thus, Figure 12.1 clearly indicates a pattern of sales penetration that can begin to offer insights into a retail development strategy (weak penetration to the south of Leeds and Bradford probably points to the most promising development opportunity in this case, given the relatively high concentration of population in this area).

The next generation of computer maps has taken capabilities for the representation of spatial pattern to new levels – see Plate 8, for example. Nevertheless, the ability to

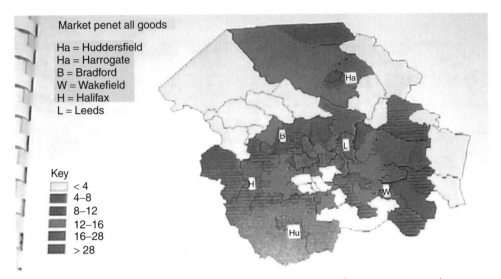

Market penet all goods

Ha = Huddersfield
Ha = Harrogate
B = Bradford
W = Wakefield
H = Halifax
L = Leeds

Key

☐ < 4
■ 4–8
■ 8–12
■ 12–16
■ 16–28
■ > 28

Figure 12.1 Market penetration map for an auto-fit operation (first generation and now very tatty!)

represent patterns in a more attractive fashion does not necessarily provide a crucial step-change in capability. Specifically, there has always been a danger with geographical information systems (GIS) that style overrides content. Again, fifteen years ago the authors were arguing about the applied weakness of GIS technologies (Birkin, Clarke, Clarke & Wilson, 1987). Much of our later research, as reported in this book and elsewhere, has served to reinforce the key point of our argument that GIS and spatial modelling represent complementary technologies; yet, at the same time, we still believe that mainstream GIS have failed to match progress in data processing and management with similar progress in spatial modelling functionality (cf. Chapter 7, Section 7.4.2).

While computer power and capabilities for representation have been accelerating so rapidly, equally important strides in the availability of data have also taken place. For example, it is now possible to get access to either a CD or web site containing the name, address and telephone numbers of the UK population for a price of about £50 (192-com, 2002). In Chapter 10, we saw that data proliferates from all kinds of different sources. One of the potential worries for professional geographers is that these changes have probably been more profoundly influential in business than academic research. With the honourable exception of the census, relatively little research has been undertaken with regard to large data sets, but at the same time, commercial organizations have built up huge customer and lifestyle databases (see Chapter 10, Section 10.4), as well as mapping and property databases such as 'Yellowpoint' and OS Landline. This trend may be seen as significant, and potentially damaging, because we do not wish to see academic research disappear completely into an intellectual cul-de-sac.

To put it another way, we do believe that the use of spatial models within commercial applications has helped enrich the discipline of geography, and also to refine and enhance the models themselves. We would point to the introduction of concepts and solutions such as the Customer Marketing Area, Idealized Representation Plan, and 'self-organized solutions' to the Representative Routing Problem as examples. We agree wholeheartedly with the sentiments of the recent 'Sainsbury Report' that observes that academia and commerce need to begin to work more closely together to promote the competitiveness of both these sectors of the UK economy (DTI, 1999, paragraphs 3.24–3.29).

It is also fair to observe that the stream of applied work, which has been described in this volume, has been promoted by a particular cultural and political environment in the UK in the late 1980s and early 1990s. In particular, the government's laissez faire attitude to economic development has been associated with a lack of interest in planning methods. It is to these kinds of planning problems that the kinds of retail and spatial interaction models described in Chapters 7 and 8 were originally directed. It is to be hoped that with a shift in emphasis of public policy back towards a certain degree of intervention for the promotion of social justice, these very models might be applied with renewed vigour within a public policy environment in the coming years.

One final question, which might usefully be addressed, is the extent to which geographical planning methods will continue to be relevant in the early years of the twenty-first century. After all, it might be argued, that the physical transfer of goods will increasingly be replaced by information flows, and traditional methods of doing business will be increasingly supplanted by e-commerce ('electronic commerce') and m-commerce (commerce by mobile telephone). There are any number of reasons why this argument does not bear close scrutiny.

1. In reality, it is (of course) the case that the vast majority of goods will still need to be distributed among a dispersed base of customers. In some cases, this may become worse. For example, if supermarket retail activity becomes dominated by home shopping, then organizations will face the problem, not simply of supplying a few hundred retail outlets from central warehouse locations, but of supplying millions of customers to their home addresses. Thus, ever greater feats of geographical organization will be required; and the development of efficient distribution techniques will provide greater cost savings and comparative advantage over the competition.

2. Many products (petrol would be a perfect example) will continue to be accessible by conventional means. Even when new methods of delivery start to become available, it is already clear that some combination of virtual and physical distribution is more likely to succeed than either one or the other (see Chapters 4 and 5). For virtual organizations such as Amazon, the challenge will be to build a physical infrastructure that supports the virtual trading environment. For established retailers, the challenge will be to restructure existing networks to provide better distribution, to be supported by new electronic trading services. Seen in these terms, it also seems likely that businesses from the 'new' and 'old' economy may be tempted to combine resources for this purpose. In this scenario, the planning models of Chapter 9, Section 9.4, would become obviously relevant.

3. In any of these future scenarios, distribution is going to become more complex, if only because of the progressive proliferation of channels. An example in relation to financial services was discussed in Chapter 10. Thus, customer behaviour will become more difficult

to predict and to manage, whereas distribution networks will become more diverse and harder to optimize.

It is our sincerely held belief, therefore, that the geographical complexity of retail activities is more likely to increase in the coming years. We look forward to many fresh research and consultancy challenges in the field of retail distribution and indeed to reporting on our progress in future volumes!

BIBLIOGRAPHY

Ahn H 2001 Applying the Balanced Scorecord concept: an experience report, Long Range Planning **34**(4): 441–461.

Alexander A 1997 Strategy and strategists: evidence from an early retail revolution in Britain *International Review of Retail, Distribution and Consumer Research* **7**(1): 61–78.

Alexander A and Pollard J 2000 Banks, grocers and the changing retailing of financial services in Britain *Journal of Retailing and Consumer Services* **7**(3): 137–147.

Alexander N 1997 *International Retailing*. Cambridge: Blackwell.

Alexander N and Mortlock W 1992 Saturation and internationalisation: the future of grocery retailing in the UK *International Journal of Retail, Distribution and Management* **20**: 33–39.

Arnold S, Handelman J, and Tigert D J 1998 The impact of a market spoiler on consumer preference structures *Journal of Retailing and Consumer Services* **58**(1): 1–13.

Batey P and Brown P J 1995 From human ecology to customer targeting: the evolution of geodemographics. In Longley P and Clarke G P (eds.) *GIS for Business and Service Planning*. Cambridge: Geoinformation, 77–103.

Batty M 1989 Urban modelling and planning: reflections, retrodictions and prescriptions. In Macmillan B (ed.) *Remodelling Geography*. Chapter 10 Oxford: Blackwell.

Beaumont J R 1991a *An Introduction to Market Analysis* CATMOG 53. Norwich: Geo-Abstracts.

Beaumont J R 1991b *GIS and market analysis*. In Maguire D, Goodchild M, and Rhind D (eds.) *Geographical Information Systems: Principles and Applications*. London: Longman, 139–151.

Belchamber J 1997 Store planning strategy – a case study from Thresher, Paper Presented to 'The Art of Store Location' Conference, Henry Stewart Conference Studies, 28/30 Little Russell Street, London, WC1A 2HN.

Bell D and Valentine G 1997 *Consuming Geographies*. London: Routledge.

Bell S 1999 Image and consumer attraction to intra-urban retail areas: an environmental psychology approach *Journal of Retailing and Consumer Services* **6**(2): 67–78.

Benoit D and Clarke G P 1997 Assessing GIS for retail location analysis *Journal of Retailing and Consumer Services* **4**(4): 239–258.

Bennett N 1998 Tesco plans petrol chain growth blitz *Sunday Telegraphy* June 28th B2.

Bennison D, Clarke I, and Pal J 1995 Locational decision making in retailing: an exploratory framework for analysis *International Review of Retail, Distribution and Consumer Research* **5**(1): 1–20.

Birkin M 1993a The simulation of whole populations with added income attributes Paper presented at the Annual Meeting of the Institute of British Geographers, Royal Holloway and Bedford New College, January.

Birkin M 1993b Geodemographics versus geolifestyles, Paper Presented to the Market Research Society, London, November 5.

Birkin M 1994 Finding the right sites: the WH Smith approach. *GIS in Business '94*. Amsterdam.

Birkin M 1995 Customer targeting, geodemographics and lifestyle approaches. In Longley P and Clarke G P (eds.) *GIS for Business and Service Planning*. Cambridge: Geoinformation, 104–149.

Birkin M, Boden P, and Williams J 2002 Spatial decision support systems for petrol forecourts. In Geertman S and Stillwell J (eds.) *Planning Support Systems in Practice*, Berlin: Springer.

Birkin M and Clarke G P 1998 GIS, geodemographics and spatial modelling: an example within the UK financial services industry *Journal of Housing Research* **9**(1): 87–112.

Birkin M, Clarke G P, Clarke M, and Wilson A G 1987 *Geographical Information Systems and urban and regional analysis: Ships in the night or the beginnings of a relationship?* Joint ESRC/NSF Meeting on 'Developments in Geographical Information Systems' Edinburgh.

Birkin M, Clarke G P, Clarke M, and Wilson A G 1994 Performance indicator applications: some British examples. In Clarke G P, Bertuglia C, and Wilson A G (eds.) *Modelling the City*. London: Routledge, 121–150.

Birkin M, Clarke G P, Clarke M, and Wilson A G 1996 *Intelligent GIS: Location Decisions and Strategic Planning*. Cambridge: Geoinformation.

Birkin M, Clarke G P, and Douglas L 2003 A model for optimizing spatial mergers *Progress in Planning* (forthcoming).

Birkin M and Clarke M 1986 Comprehensive dynamic urban models: integrating macro and micro approaches. In Griffith D A and Haining R P (eds.) *Transformations Through Space and Time*. Dordrecht: Martinus Nijhoff, 165–191.

Birkin M and Clarke M 1988 SYNTHESIS: A SYNTHetic Spatial Information System for urban modelling and spatial planning *Environment and Planning A* **20**: 1645–1671.

Birkin M and Clarke M 1989 The generation of individual and household incomes at the small area level using Synthesis *Regional Studies* **23**: 535–548.

Birkin M, Clarke M, and George F 1995 The use of parallel computers to solve non-linear spatial optimisation problems: an application to network planning *Environment and Planning A* **27**: 1049–1068.

Birkin M and Foulger F 1992 Sales performance and sales forecasting using spatial interaction modelling: the WH Smith approach, working paper 92/91, School of Geography, University of Leeds.

Blalock H M 1974 *Social Statistics*. (second edition). Tokyo: Mcgraw-Hill Kogakusha.

Bodkin C D and Lord J D 1997 Attraction of power centres *International Review of Retail, Distribution and Consumer Research* **7**(2): 93–108.

Bond S 1997 Gravity modelling and its applicability to the internationalisation of business, Paper Presented to 'The Art of Store Location' Conference, Henry Stewart Conference Studies, 28/30 Little Russell Street, London, WC1A 2HN.

Borchert J G 1995 Retail planning policy in the Netherlands. In Davies R L (ed.) *Retail Planning Policies in Western Europe*. London: Routledge, 160–181.

Boyce D, Day N, and McDonald C 1970 Metropolitan Plan-Making, Monograph Series Number 4, Regional Science Research Institute, Philadelphia, Pa.

Bramley G and Lancaster S 1998 Modelling local and small-area income distributions in Scotland *Environment and Planning C* **16**(6): 681–706.

Brown P J B 1991 Exploring geodemographics. In Masser I and Blakemore M (eds.) *Handling Geographical Information*. Harlow: Longman.

Burt S 1995a Retail internationalization: evolution of theory and practice. In McGoldrick P J and Davies G (eds.) *International Retailing: Trends and Strategies*. London: Pitman Publishing, 51–73.

Burt S 1995b Carrefour: internationalizing innovation. In McGoldrick P J (ed.) *Cases in Retail Management*. London: Pitman Publishing, 154–164.

Burt S and Limmack R 2001 Takeovers and shareholder returns in the retail industry *International Review of Retail, Distribution and Consumer Research* **11**(1): 1–22.

Burt S and Sparks L 1994 Structural change in grocery retailing in Great Britain: a discount reorientation? *International Review of Retail, Distribution and Consumer Research* **4**: 195–217.

Burt S and Sparks L 1995 Understanding the arrival of limited-line discount stores in Britain *European Management Journal* **13**: 110–119.

Burt S and Sparks L 2001 The implications of Wal-Mart's takeover of ASDA *Environment and Planning A* **33**: 1463–1487.

CACI 1993 CACI's insite system in action *Marketing Systems Today* **8**(1): 10–13.

CACI 2001 Corporate web-site, *www.caci.co.uk*

Cairncross F 1997 *The Death of Distance: How the Communications Revolution will Change Our Lives*. London: Orion Books.

Cathelat B 1991 *Panorama des styles de vie*. Paris: Les Editions d'Organisation.

Cathelat B 1993 *Socio-Styles*. London: Kogan Page.

Cavanagh E 1999 Fashion retailers target the UK *European Retail Digest* **March**: 75–76.

Claritas 1996 PRIZM. *Transforming Market Segmentation* London: Claritas.

Claritas 2001 Corporate web-site. *www.claritas.co.uk*.

Clarke D B 1996 The limits to retail capital. In Wrigley N and Lowe M (eds.) *Retailing, Consumption and Capital: Towards the New Retail Geography*. London: Longman, 284–301.

Clarke G P (ed.) 1996 *Microsimulation for Urban and Regional Policy Analysis*. London: Pion.

Clarke G P 1999 'Geodemographics, marketing and retail location'. In Pacione M (ed.) *Applied Geography*. London: Routledge, 577–592.

Clarke G P and Clarke M 2001 Applied spatial interaction modeling. In Clarke G P and Madden M (eds.) *Regional Science in Business*. Berlin: Springer.

Clarke G P, Eyre H, and Guy C 2002 Deriving indicators of access to food retail provision in British cities: studies of Cardiff. Leeds and Bradford. *Urban Studies*. In press.

Clarke G P, Langley R, and Cardwell W 1998 Empirical applications of dynamic spatial interaction models Computers *Environment and Urban Systems* **22**(2): 157–184.

Clarke G P and Madden M (eds.) 2001 *Regional Science in Business*. Berlin: Springer.

Clarke G P and Stillwell J C H (eds.) 2002 *Applied GIS and Spatial Modelling*. Chichester: John Wiley & Sons.

Clarke G P and Wilson A G 1994 A new geography of performance indicators for urban planning. In Bertuglia C S, Clarke G P, and Wilson A G *Modelling the City: Performance, Policy and Planning*. London: Routledge, 55–81.

Clarke I 2000 Retail power, competition and local consumer choice in the UK grocery sector *European Journal of Marketing* **8**: 975–1002.

Clarke I, Bennison D, and Guy C 1994 The dynamics of UK grocery retailing at the local scale *International Journal of Retail, Distribution and Management* **22**(6): 11–20.

Clarke I, Horita M, and Mackaness W 2000 The spatial knowledge of retail decision makers: capturing and interpreting group insight using a composite cognitive map *International Review of Retail Distribution and Consumer Research* **10**(3): 265–285.

Clarke M and Wilson A G 1983 The dynamics of urban spatial structure: progress and problems *Journal of Regional Science* **21**: 1–18.

Clarke M and Wilson A G 1985 The dynamics of urban spatial structure: the progress of a research programme *Transactions of the Institute of British Geographers* **10**: 427–451.

Clarkson R M, Clarke-Hill C M, and Robinson T 1996 UK supermarket location assessment *International Journal of Retail and Distribution Management* **24**(6): 22–33.

Cliquet G 2000a Large format retailers: a French tradition despite reactions *Journal of Retailing and Consumer Services* **7**(4): 183–196.

Cliquet G 2000b Plural forms in store networks: a model for store network evolution *International Review of Retail, Distribution and Consumer Research* **10**(4): 369–387.

Coombes M G, Green A E, and Openshaw S 1986 An efficient algorithm to generate official statistical reporting areas: the case of the 1984 travel-to-work areas revision in Britain *Journal of the Operational Research Society* **37**(10): 943–953.

Cooper L 1972 The transportation-location problem *Operations Research* **11**: 331–343.

Cope N 1996 *Retail in the Digital Age*. London: Bowerdean.

Corrigan P 1997 *The Sociology of Consumption*. London: Sage.

Crewe L 2000 Geographies of retailing and consumption *Progress in Human Geography* **24**(2): 275–290.

Crewe L and Davenport E 1992 The puppet show: changing buyer-supplier relationships within clothing retailing *Transactions of the Institute of British Geographers* **17**: 183–197.

Crewe L and Gregson N 1998 Tales of the unexpected: exploring car boot sales as marginal spaces of contemporary consumption *Transactions of the Institute of British Geographers* **23**: 39–53.

Crewe L and Lowe M 1995 Gap on the map? Towards a geography of consumption and identity *Environment and Planning A* **27**: 1877–1898.

Croft M J 1994 *Market Segmentation: A Step-by-Step Guide to Profitable New Business*. London: Routledge.

Davies R L 1984 *Retail and Commercial Planning*. Beckenham: Croom Helm.

Davies R L and Rogers D S 1984 *Store Location and Store Assessment Research*. Chichester: John Wiley & Sons.

Davis L 1991 *Handbook of Genetic Algorithms*. New York: Van Nostrand Reinhold.

Dawson J A (ed.) 1981 *Retail Geography*. London: Croom Helm.

Dawson J A 1993 The internationalization of retailing. In Bromley R and Thomas C (eds.) *Retail Change: Contemporary Issues*. London: UCL Press, 15–40.

Dawson J A 1995 Retail change in the European Community. In Davies R L (ed.) *Retail Planning Policies in Western Europe*. London: Routledge, 1–30.

Dawson J A 2000 Retailing at century end: some challenges for management and research *International Review of Retail, Distribution and Consumer Research* **10**(2): 119–148.

De Kare-Silver M 2000 *E-shock: e-strategies for Retailers and Manufacturers*. Basingstoke: Palgrove.

Debenham J, Clarke G P, and Stillwell J 2001 Improving geodemographics for business: adding supply-side variables, working paper, School of Geography, University of Leeds, LS2 9JT.

Diplock G 1998 Building new spatial interaction models by using genetic programming and a super computer *Environment and Planning A* **30**(10): 1893–1904.

Docherty A M 1999 Explaining international retailer's market entry mode strategy: internalization theory, agency theory and the importance of information asymmetry *International Review of Retail, Distribution and Consumer Research* **9**(4): 379–402.

Doz Y L and Hamel G 1998 *Alliance Advantages*. Harvard, Boston: Harvard Business School.

DTI 1999 Biotechnology Clusters: Report of a team led by Lord Sainsbury, Minister for Science. HMSO, London.

Duke R 1989 A structural analysis of the UK grocery retail market *British Food Journal* **91**(5): 17–22.

Duke R 1998 Buyer-supplier relationships in UK food retailing *Journal of Retailing and Consumer Services* **5**(2): 93–104.

Experian 2001 *Pixel: High Resolution Customer Segmentation*. Nottingham, UK: Experian.

Fagan M 1999 Northern exposure, *Sunday Telegraph*, 4th April, B3.

Feng Z and Flowerdew R 1997 Fuzzy geodemographics: a contribution from fuzzy clustering methods. In Carver S (ed.) *Innovations in GIS 5*. London: Taylor & Francis, 119–127.

Fenwick I 1978 *Techniques in Store Location Research: A Review and Applications*. Corbridge: Retail and Planning Associates.

Fernie J, Moore C, Lawrie A, and Hallsworth A 1997 The internationalization of the high fashion brand: the case of central London *Journal of Product and Brand Management* **6**(3): 151–162.

Fernie J and Sparks L (eds.) 1998 *Logistics and Retail Management*. London: Kogan Page.

Fernie J and Staines H 2001 Towards an understanding of European grocery supply chains *Journal of Retailing and Consumer Services* **8**(1): 29–36.

Field C 1997 *The Future of the Store: New Formats and Channels for a Changing Retail Environment*. London: Financial Times Business.

Fik T J and Mulligan G F 1990 Spatial flows and competing central places: towards a general theory of hierarchical interaction *Environment and Planning A* **22**: 527–549.

Financial Times 1998 Why maps are being redrawn, *Financial Times*, May 11.

Fischer M M and Getis A 1999 New advances in spatial interaction theory *Papers in Regional Science* **78**(2): 117–118.

Fischer M M, Hlavackova-Schindler K, and Reismann M 1999 A global search procedure for parameter estimation in neural net spatial interaction modeling *Papers in Regional Science* **78**(2): 119–134.

Foord J, Bowlby S, and Tillsley C 1996 The changing place of retailer-supplier relations in British retailing. In Wrigley N and Lowe M (eds.) *Retailing, Consumption and Capital: Towards the New Retail Geography*. London: Longman, 68–89.

Fotheringham A S 1983 A new set of spatial interaction models: the theory of competing destinations *Environment and Planning A* **15**(1): 15–36.

Fotheringham A S 1986 Modelling hierarchical destination choice *Environment and Planning A* **18**: 401–418.

Fotheringham A S 1988 Consumer store choice and choice set definition *Marketing Science* **7**(3): 299–310.

Fotheringham A S, Brunsdon C, and Charlton M 2000 *Quantitative Geography: Perspectives on Spatial Data Analysis*. London: Sage.

Fotheringham A S and Knudson D C 1986 Modeling discontinuous change in retailing systems: extensions of the Harris-Wilson framework with results from a simulated urban retailing system *Geographical Analysis* **18**(4): 295–312.

Fotheringham A S, and O'Kelly M 1989 *Spatial Interaction Models: Formulation and Applications*. Dordrecht: Kluwer.

Freathy P and O'Connell F 1998 *European Airport Retailing: Growth Strategies for the New Millennium*. Hampshire: MacMillan.

George F, Radcliffe N, Smith M, Birkin M, and Clarke M 1997 Algorithms for solving a spatial optimisation problem on a parallel computer *Concurrency: Practice and Experience* **9**: 753–780.

Geertman S and Stillwell J C H 2002 *Decision Support Systems*, Berlin: Springer.

Ghosh A and Harche F 1993 Location-allocation models in the private sector: progress, problems and prospects *Locational Science* 1: 81–106.

Ghosh A and MacLafferty S 1987 *Locational Strategies for Retail and Service Firms*. Lexington MA: Lexington Books.

GMAP 2000 *The GMAP Channel Usage Survey*. Leeds: GMAP.

GMAP 2001 Corporate web-site, *www.gmap.com*.

Goldman A 2000 Supermarkets in China; the case of Shanghai *International Review of Retail, Distribution and Consumer Research* **10**(1): 1–22.

Goodwin D R and McElwee R E 1999 Grocery shopping and an ageing population, a research note *International Review of Retail, Distribution and Consumer Research* **9**(4): 403–409.

Goss J 1996 Disquiet on the waterfront: reflections on nostalgia and utopia in the urban archetypes of festival market places *Urban Geography* **17**: 221–247.

Graff T O 1998 The locations of Wal-Mart and Kmart Supercenters: contrasting corporate strategies *Professional Geographer* **50**(1): 46–57.

Graff T O and Ashton D 1993 Spatial diffusion of Wal-Mart: contagious and reverse hierarchical elements *Professional Geographer* **46**: 19–29.

Green A 1998 Geography of earnings and income in the 1990s *Environment and Planning C* **16**: 632–647.

Gregson N and Crewe L 1997 The bargain, the knowledge and the spectacle: making sense of consumption in the space of the car-boot sale *Environment and Planning D* **15**: 87–112.

Guldmann J M 1999 Competing destinations and intervening opportunities interaction models of inter-city telecommunication flows *Papers in Regional Science* **78**(2): 179–194.

Guy C 1980 *Retail Location and Retail Planning*. Farnborough: Gower.

Guy C 1994a Grocery store saturation: has it arrived yet? *International Journal of Retail and Distribution Management* **22**(1): 3–11.

Guy C 1994b *The Retail Development Process*. London: Routledge.

Guy C 1995 Retail store development at the margin *Journal of Retailing and Consumer Services* **2**: 25–32.

Guy C 1996a Grocery store saturation in the UK – the continuing debate *International Journal of Retail and Distribution Management* **24**(6): 3–10.

Guy C 1996b Corporate strategies in food retailing and their local impacts: a case study of Cardiff *Environment and Planning A* **28**: 1575–1602.

Guy C 1998a Alternative use valuation, open A1 planning consent, and the development of retail parks *Environment and Planning A* **30**: 37–47.

Guy C 1998b Controlling new retail spaces: the impress of planning policies in Western Europe *Urban Studies* **35**: 953–979.

Guy C 2000 From crinkly sheds to fashion parks: the role of financial investment in the transformation of retail parks *International Review of Retail, Distribution and Consumer Research* **10**(4): 389–400.

Hahn B 2000 Power centres: a new retail format in the USA *Journal of Retailing and Consumer Services* **7**(4): 223–232.

Hallsworth A 1990 The lure of the USA *Environment and Planning A* **22**: 551–557.

Hallsworth A 1997 Rethinking retail theory: circuits of power as an integrative paradigm *Geographical Analysis* **29**(4): 329–338.

Hamnett C 1996 Social polarization, economic restructuring and the welfare state *Urban Studies* **33**: 1407–1431.

Harbison J and Pekar T Jr. 1998 *Smart Alliances*. San Francisco: Jossey-Bass.

Harris B and Wilson A G 1978 Equilibrium values and dynamics of attractiveness terms in production-constrained spatial interaction models *Environment and Planning A* **10**: 371–388.

Healy & Baker 1998 *New Trends in European Retailing: Retail Warehousing*. London: Healy & Baker.

Hernandez J A 1998 The role of GIS within retail location decision making, unpublished PhD thesis, Department of Retailing, Manchester Metropolitan University, Manchester.

Hernandez J A, Bennison D, and Cornelius S 1999 The organisational context of retail locational planning *GeoJournal* **45**(4): 299–308.

Hernandez J A and Bennison D 2000 The art and science of retail location decisions *International Journal of Retail, Distribution and Management* **28**(8/9): 357–367.

Hodgson M J 1978 Towards more realistic allocation in location-allocation models: an interaction approach *Environment and Planning A* **10**: 1273–1285.

Huff D L 1963 A probabilistic analysis of shopping center trade areas *Land Economics* **39**: 81–90.

Holland J H 1975 *Adaptation in Natural and Artificial Systems*. Ann Arbor: University of Michigan Press.

Hoschka T C 1993 *Cross-Border Entry in European Retail Financial Services: Determinants, Regulation and the Impact on Competition*. London: MacMillan.

Hubbard N 1999 *Acquisition*. Basingstoke: MacMillan.

Hughes A 1996 Forging new cultures of food retailer-manufacturer relations. In Wrigley N and Lowe M (eds.) *Retailing, Consumption and Capital: Towards the New Retail Geography*. London: Longman, 90–115.

Hughes A 1999 Constructing competitive spaces: on the corporate practice of British retailer-supplier relationships *Environment and Planning A* **28**: 2201–2226.

Hurley S and Moutinho L 1996 Approximate algorithms for marketing management problems *Journal of Retailing and Consumer Services* **3**: 145–154.

Integral Solutions 1997 Halfords and Clementine: state of the art in store location, Fact sheet in 'The Art of Store Location' Conference, Henry Stewart Conference Studies, 28/30 Little Russell Street, London, WC1A 2HN.

Ireland P 1994 GIS: another sword for St. Michael *Mapping Awareness* **April**: 26–29.

Johnston R J 2000a Districting algorithm. In Johnston R J, Gregory D, Pratt G, and Watts M (eds.) *The Dictionary of Human Geography.* (fourth edition). Oxford: Blackwell.

Johnston R J 2000b Malapportionment. In Johnston R J, Gregory D, Pratt G, and Watts M (eds.) *The Dictionary of Human Geography.* (fourth edition). Oxford: Blackwell.

Jones G R J 1976 Multiple estates and early settlement. In Sawyer P H (ed.) *Medieval Settlement: Continuity and Change.* London: Edward Arnold.

Jones G R J 1985 Multiple estates perceived *Journal of Historical Geography* **11**: 352–363.

Jones Lang Wootton and Oxford Institute of Retail Management 1998 *Shopping for New Markets: Retail Opportunities in Central Europe.* London: Jones Long Wootton.

Jones K and Biasiotto M 1999 Internet retailing: current hype or future reality? *International Review of Retail, Distribution and Consumer Research* **9**(1): 69–79.

Jones K and Doucat M 2000 Big-box retailing and the urban retail structure: the case of the Toronto area *Journal of Retailing and Consumer Services* **7**(4): 233–247.

Jones K and Hernandez T 2002 Retail applications of spatial modelling. In Clarke G P and Stillwell J C H (eds.) *Applied GIS and Spatial Modelling.* Chichester: John Wiley & Sons, (forthcoming).

Jones K and Simmons J 1987 *Location, Location, Location.* London: Methuen.

Jones K and Simmons J 1990 *The Retail Environment.* London: Routledge.

Jones P M 1982 Hypermarkets and superstores–saturation or future growth? *Retail and Distribution Management* **10**(4): 20–27.

Kaplan C 1995 A world without boundaries: the Body shop's trans/national geographies *Social Text* **43**: 45–66.

Kohonen T 1984 *Self-Organisation and Associative Memory.* Berlin: Springer-Verlag.

Kantorvich Y 1992 Equilibrium models of spatial interaction with location capacity constraints *Environment and Planning A* **24**: 1071–1095.

Knudsen D and Fotheringham A S 1986 Matrix comparison, goodness-of-fit, and spatial interaction modelling *International Regional Science Review* **10**(2): 127–47.

KPMG 1997 *Colouring the Map: Mergers and Acquisitions in Europe.* London: KPMG Management Group.

Laaksonen H and Reynolds J 1994 Own branding in food retailing across Europe *Journal of Brand Management* **2**(1): 37–46.

Lamey J 1997 *Retail Internationalization.* London: Financial Times Business.

Lamont M and Molnar V G 2001 How blacks use consumption to shape their collective identity *Journal of Consumer Culture* **1**(1): 31–45.

Langston P, Clarke G P, and Clarke D B 1997 Retail saturation, retail location and retail competition: an analysis of British food retailing *Environment and Planning A* **29**: 77–104.

Langston P, Clarke G P, and Clarke D B 1998 Retail saturation: the debate in the mid 1990s *Environment and Planning A* **30**: 49–66.

Laulajainen R 1987 *Spatial Strategies in Retailing.* Dordrecht: Reidal.

Laulajainen R 1988 The spatial dimensions of an acquisition *Economic Geography* **64**: 170–187.

Laulajainen R 1990 Defense by expansion: the case of Marshall Field. *The Professional Geographer* **42**: 277–288.

Lawson D 1996 Boys' own story *Property Week* 29th November 30–31.

Lee D B 1973 Requiem for large scale models *Journal of the American Institute of Planners* **39**: 163–178.

Leefland P and Van Raaji W 1995 The changing consumer in the European Union *International Journal of Research in Marketing* **12**: 373–387.

Leyshon A and Thrift N J 1995 Geographies of financial exclusion: financial abandonment in Britain and the United States *Transactions of the Institute of British Geographers* **20**(3): 312–341.

Leyshon A and Thrift NJ 1999 Lists come alive: electronic systems of knowledge and the rise of credit-scoring in retail banking, Economy and Society, **28**(3), 434–466.

Longley P, Goodchild M, Maguire D, and Rhind D 2001 *Geographical Information and Science*. London: John Wiley & Sons.

Lord J D 2000 Retail saturation: inevitable or irrelevant? *Urban Geography* **21**(4): 342–360.

Lord J D and Lynds C 1981 The use of regression models in store location research *Akron Business and Economic Review* **10**: 13–19.

Lowe M 2000 Britain's regional shopping centres: new urban forms? *Urban Studies* **37**: 261–274.

Lowe M and Wrigley N 2000 Retail and the urban *Urban Geography* **21**(7): 640–653.

Lynch R 1993 *Cases in European Marketing*. London: Kogan Page.

Mahoney R 1989 Should local authorities use a corporate or departmental GIS? *Mapping Awareness* **3**(2): 57–59.

Markham J 1998 *The Future of Shopping: Traditional Patterns and Net Effects*. Basingstoke: MacMillan Business Press.

Mason J B and Meyer J L 1981 *Modern Retailing: Theory and Practice*. Texas: Business Publications.

McGoldrick P J and Davies G (eds.) 1995 *International Retailing: Trends and Strategies*. London: Pitman.

McKinnon A C 1986 The physical distribution strategies of multiple retailers *International Journal of Retailing* **1**(2): 49–63.

McKinnon A C 1989 *Physical Distribution Systems*. London: Routledge.

Mendelsohn M 1992a *Franchising in Europe*. London: Cassell.

Mendelsohn M 1992b *The Guide to Franchising*. London: Cassell.

Metton A 1995 Retail planning policy in France. In Davies R L (ed.) *Retail Planning Policies in Western Europe*. London: Routledge, 62–77.

Mills L 1999 Asda plans £100 m supercentre chain *Sunday Telegraph* April 4th, B9.

Moore A T and Attewell G 1991 To be and where not to be: the Tesco approach to locational analysis *Operational Research Insight* **4**: 21–24.

Morgan M S and Chintagunta P K 1997 Forecasting restaurant sales using self-selectivity models *Journal of Retailing and Consumer Services* **4**(2): 117–128.

Morris C 1993 Lifestyle data. In Leventhal B, Moy C, and Griffin J (eds.) *An Introductory Guide to the 1991 Census*. London: The Market Research Society/ NTC Publications.

Myers W 1993 *Food Retail Perception and Reality*. 32 St. Mary's Hill, London: Henderson Crosthwaite Institutional Brokers, EC3R 1GG.

National Demographics and Lifestyles 1993 *The Lifestyle Selector*, Product Brochure, London: NDL.

National Demographics and Lifestyles 1993 *Targeting for the car industry*, Product Brochure, London: NDL.

National Retail Planning Forum 1998 *Shopping Centres in the Pipeline*. London: National Retail Planning Forum.

Nayga R M Jr. and Weinberg Z 1999 Supermarket access in the inner cities *Journal of Retailing and Consumer Services* **6**(3): 141–145.

Nooteboom B 1999 *Inter-Firm Alliances*. London: Routledge.

Norris S 1999 Factory outlet centres into the New Millennium: 'fact and fiction' *European Retail Digest* **March**: 41–43.

Ogbonna E and Wilkinson B 1990 Corporate strategy and corporate culture: the view from the checkout on the management of change in the UK supermarket industry *Personnel Review* **19**(3): 9–15.

Ogbonna E and Wilkinson B 1998 Power relations in the UK grocery supply chain *Journal of Retailing and Consumer Services* **5**(2): 77–86.

O'Grady S and Lane H W 1997 Culture: an unnoticed barrier to Canadian retail performance in the USA *Journal of Retailing and Consumer Services* **4**(3): 159–170.

O'Kelly M 2001 Retail market share and saturation *Journal of Retailing and Consumer Services* **8**(1): 37–46.

Openshaw S 1977 A geographical solution to scale and aggregation problems in region-building, partitioning and spatial modelling *Transactions of the Institute of British Geographers* **2**: 459–472.

Openshaw S 1993 Modelling spatial interaction using a neural net. In Fischer M M and Nijkamp P (eds.) *GIS, Spatial Modelling and Policy Evaluation*. Berlin, Germany: Springer, 147–164.

Openshaw, S 1998 In Longley P A, Brooks S M, and Mcdonnell B (eds.) *Building Automated Geographical Analysis and Exploration Machines in Geocomputation: A Primer*. Chichester: Wiley, 95–115.

Openshaw S and Openshaw C 1997 *Artificial Intelligence in Geography*. John Wiley & Sons.

Openshaw S, Wymer C, and Blake M 1995 Using neurocomputing methods to classify Britain's residential areas. In Taylor P (ed.) *Innovations in GIS 2*. London: Taylor & Francis, 97–111.

Oppewal H and Timmermans H 2001 Discrete choice modelling: basic principles and application to parking policy assessment. In Clarke G P and Madden M (eds.) *Regional Science in Business*. Berlin: Springer, 97–114.

Orcutt G, Merz J, and Quinke H 1986 *Microanalytic Simulation Models to Support Social and Financial Policy*. Amsterdam: North Holland.

Pacione M 1974 Measures of the attraction factor *Area* **6**: 279–282.

Pal J, Bennison D, Clarke I, and Byrom J 2001 Power, policy networks and planning *International Review of Retail, Distribution and Consumer Research* **11**(3): 225–246.

Palmer T and Beddall C 1997 The rise and rise of Tesco *The Grocer* September 20: 40–44.

Pellegrini L 1995 Retail planning policy in Italy. In Davies R L (ed.), *Retail Planning Policies in Western Europe* London: Routledge, 144–159.

Peter Shearman Associates 1996 *A Tale of Three Cities: What Makes a Successful City Centre?* London: National Retail Planning Forum.

Phiri P 1980 Calculation of the equilibrium configuration of shopping facility sizes *Environment and Planning A* **12**: 983–1000.

Piacentini M, Hibbert S, and Al-Dajani H 2001 Diversity in deprivation: exploring the grocery shopping behaviour of disadvantaged consumers *International Review of Retail, Distribution and Consumer Research* **11**(2): 141–158.

Pike R and Neale B 1993 *Corporate Finance and Investment: Decisions and Strategies.* Englewood Cliffs, NJ: Prentice Hall.

Pinch S 1993 Social polarization: a comparison of evidence from Britain and the USA *Environment and Planning A* **25**: 779–795.

Poole R, Clarke G P, and Clarke D B 2002a Growth, concentration and regulation in European grocery retailing *European Urban and Regional Studies* **9**(2): 167–186.

Poole R, Clarke G P, and Clarke D B 2002b Food retailers and regional monopolies *Regional Studies* (forthcoming).

Poyner M 1987 The changing consumer. In McFadyen E (ed.) *The Changing Face of British Retailing.* London: Newman Books, 94–103.

Press W *et al.* 1989 *Numerical Recipes,* Cambridge: Cambridge University Press.

Quadstone 2000 Database Marketing (Seminar), Institute of Directors, London.

Quinn B 1998 Towards a framework for the study of franchising as an operating mode for international retail companies *International Review of Retail, Distribution and Consumer Research* **8**(4): 445–467.

Ravenstein E G 1885 The laws of immigration *Journal of the Royal Statistical Society* **48**: 167–235 and 241–305.

Reece D 1996 Phone deals smash branches, Sunday Telegraph, August 18th, Finance section, **7**.

Reilly W J 1931 *The Law of Retail Gravitation.* New York: GP Putnam & Sons.

Reynolds J 1991 GIS for competitive advantage: the UK retail sector *Mapping Awareness* **5**(1): 33–36.

Reynolds J 1993 The role of GIS in European cross-border retailing *Mapping Awareness* **7**(2): 20–25.

Reynolds J 1998 Opportunities for electronic commerce *European Retail Digest* **March**: 5–9.

Rijk F J A and Vorst A C F 1983a Equilibrium points in an urban retail model and their connection with dynamic systems *Regional Science and Urban Economics* **13**: 383–399.

Rijk F J A and Vorst A C F 1983b On the uniqueness and existence of equilibrium points in an urban retail model *Environment and Planning A* **15**: 475–482.

Robertet E 1997 How social change affects retail habits: a typology of the European population *European Retail Digest* **Winter**: 4–14.

Rogers D S and Green H L 1979 A new perspective on forecasting store sales: applying statistical models and techniques in the analog approach *Geographical Review* **69**: 449–458.

Roy J R 1999 Areas, nodes and networks: some analytical considerations *Papers in Regional Science* **78**(2): 135–156.

Rushton G, Goodchild M F, and Ostresh L M (eds.) 1972 *Computer programmes for location-allocation problems,* Monograph No. 6, Department of Geography, University of Iowa.

Sampson P 1992 People are people the world over *Marketing and Research Today* **20**: 236–244.

Sayer R A 1976 *A Critique of Urban Modelling*. Oxford: Pergamon Press.

Scholten H and Meijer E 2002 Spatial retail information: making it more simple. In Clarke G P and Stillwell J C H (eds.) *Applied GIS and Spatial Modelling*. Chichester: John Wiley & Sons. (forthcoming)

See L and Openshaw S 2001 Fuzzy geodemographic targeting. In Clarke G P and Madden M (eds.) *Regional Science in Business*. Berlin: Springer.

Sen A and Smith T E 1995 *Gravity Models of Spatial Interaction Behaviour*. Berlin: Springer.

Shackleton R 1998 Part-time working in the 'super-service' era: labour force restructuring in the UK food retailing industry during the late 1980s, early 1990s *Journal of Retailing and Consumer Services* **5**(4): 223–234.

Shaw G 1992 The evolution and impact of large-scale retailing in Britain. In Benson J and Shaw G (eds.) *The Evolution of Retail Systems*. Leicester: University Press.

Silcock L, Clarke G P, Clarke D, and Wrigley N 1999 Grocery provision in the USA: room for expansion? *International Journal of Retail, Distribution and Management* **27**(1): 8–21.

Simkin L P 1990 Evaluating a store location *International Journal of Retail and Distribution Management* **18**(4): 33–38.

Sleight P 1993 *Targeting Customers: How to Use Geodemographic and Lifestyle Data in Your Business*. Henley on Thames, London: NTC Publications.

Smith D 1998 Logistics in Tesco: past, present and future. In Fernie J and Sparks L (eds.) *Logistics and Retail Management*. London: Kogan Page, 154–183.

Sparks L 1990 Spatial-structural relationships in retail corporate growth: a case study of kwik save group PLC *Service Industries Journal* **10**: 25–84.

Sparks L 1995 Reciprocal retail internationalization: the Southland corporation, Ito-Yokado and 7-eleven convenience stores *The Service Industries Journal* **15**(4): 57–96.

Sparks L 1996a Space wars: Wm low and the 'auld enemy' *Environment and Planning A* **28**: 1465–1484.

Sparks L 1996b Challenge and change: Shoprite and the restructuring of grocery retailing in Scotland *Environment and Planning A* **28**: 261–284.

Sparks L 1998 The retail logistics transformation. In Fernie J and Sparks L (eds.) *Logistics and Retail Management*. London: Kogan Page.

Spencer A H 1978 Deriving measures of attractiveness for shopping centres *Regional Studies* **12**: 713–726.

SPSS 1995 *A Statistical Package for the Social Sciences. Users Guide*. Chicago: SPSS.

SPSS 2002 Corporate web-site, *www.spss.com*.

Sternquist B 1998 *International Retailing*. London: Fairchild Books.

Sudarsanam P S 1995 *Mergers and Acquisitions*. London: Prentice-Hall.

Talluri S 2000 A benchmarking method for business process reengineering and improvement *International Journal of Flexible Manufacturing Systems* **12**(4): 291–304.

Tewdwr-Jones M 1997 Plans, policies and inter-governmental relations: assessing the role of national planning guidance in England and Wales *Urban Studies* **34**: 141–162.

Thrall G I, Casey J, and Quintana, A 2001 Geopsychographic lifestyle segmentation providers *Geospatial Solutions* **11** (4):

Timmermans H 1981 Multi-attribute shopping models and ridge regression models *Environment & Planning A* **13**: 43–56.

Treadgold A 1991 *Managing International Retail Business*. Oxford: Oxford Institute for Retail Management.

Trefson H 1994 Restructuring branch networks in the changing financial world: the meaning of GIS for retail banks *European Retail Marketing* **Spring**: 17–22.

University of Edinburgh 2001 About Great Britain Mosaic, University of Edinburgh Data Library, *http://datalib.ed.ac.uk/EUDL/appendixc.html*.

Van Kenhowe P and De Wulf K 2000 Income and time pressure: a person-situation grocery retail typology *International Review of Retail, Distribution and Consumer Research* **10**(2): 149–166.

Vielberth H 1995 Retail planning policy in Germany. In Davies R L (ed.) *Retail Planning Policies in Western Europe.* London: Routledge, 78–103.

Wagner S 1997 *Understanding Green Consumer Behaviour; A Qualitative Cognitive Approach.* London: Routledge.

White R 1995 *Cross-Border Retailing.* London: Financial Times.

Whittam G and Clarke G P 2002 Doing business in emerging markets, working paper, School of Geography, University of Leeds, (forthcoming).

Whysall P 1997 Interwar retail internationalisation: Boots under American ownership *International Review of Retail, Distribution and Consumer Research* **7**(2): 157–170.

Williams C C and Windebank J 2000 Modes of good acquisition in deprived neighbourhoods *International Review of Retail, Distribution and Consumer Research* **10**(1): 1–22.

Williams H C 1981 Random utility theory and probabilistic choice models In Wilson A G, Coelho J, Macgill S, and Williams H C (eds.) *Optimisation in Location and Transport Analysis* Chichester: Wiley.

Williams P and Hubbard P 2001 Who is disadvantaged? Retail change and social exclusion *International Review of Retail, Distribution and Consumer Research* **11**(3): 267–286.

Williamson P, Birkin M, and Rees P 1998 The simulation of whole populations using data from small area statistics and samples of anonymised records *Environment and Planning A* **30**: 785–816.

Wilson A G 1967 *A statistical theory of spatial distribution models Transportation Research* **1**: 253–269.

Wilson A G 1974 *Models for Urban and Regional Planning.* Chichester: John Wiley & Sons.

Wilson A G 1981 *Catastrophe Theory and Bifurcation: Applications to Urban and Regional Systems.* London: Croom Helm.

Wilson A G 1983 A generalised and unified approach to the modelling of service supply structures (Working Paper 352). Leeds, UK: School of Geography, University of Leeds.

Wrigley N (ed.) 1988 *Store Choice, Store Location and Market Analysis.* London: Routledge.

Wrigley N 1989 The lure of the USA: further reflections on the internationalization of British grocery retail capital *Environment and Planning A* **21**: 283–288.

Wrigley N 1992 *Antitrust regulation and the restructuring of grocery retailing in Britain and the USA Environment and Planning A* **24**: 727–749.

Wrigley N 1994 After the store wars: towards a new era of competition in UK food retailing *Journal of Retailing and Consumer Services* **1**: 5–20.

Wrigley N 1996 Sunk costs and corporate restructuring: British food retailing and the property crisis. In Wrigley N and Lowe M (eds.) *Retailing, Consumption and Capital: Towards the New Retail Geography.* London: Longman, 116–136.

Wrigley N 1997 Foreign retail capital on the battlefields of Connecticut: competition at the local scale and its implications *Environment and Planning A* **29**: 1141–1152.

Wrigley N 1998 Understanding store development programmes in post-property crisis UK food retailing *Environment and Planning A* **30**: 15–35.

Wrigley N 1999 Market rules and spatial outcomes; insights from the corporate restructuring of US food retailing *Geographical Analysis* **31**: 288–309.

Wrigley N 2000 The globalisation of retail capital. In Clark G L, Feldman M, and Gertler M S (eds.) *The Oxford Handbook of Economic Geography*. Oxford: Oxford University Press, 292–313.

Wrigley N 2001 Transforming the corporate landscape of US food retailing: market power, financial re-engineering and regulation *Tijdschrift voor Economische en Sociale Geografie* (forthcoming).

Wrigley N and Clarke G P 1999 Discount shake-out: the transformation of UK discount food retailing, 1993–1998, unpublished paper, available from 1st author, Department of Geography, University of Southampton, Southampton, SO17 1BJ.

Wrigley N and Lowe M 1996 (eds.) *Retailing, Consumption and Capital*. London: Longman.

Wrigley N and Lowe M 2002 *Reading Retail: A Geographical Perspective on Retailing and Consumption Spaces*. London: Arnold.

Wrigley N, Lowe M and Currah A 2002 Retailing and e-tailing *Urban Geography* **23**(2): (forthcoming).

Wrigley N, Warm D, Margetts B, and Whelan A 2002 'Assessing the impact of improved retail access on diet in a 'Food Desert': A preliminary report' *Urban Studies* (forthcoming).

Zadeh L A 1965 Fuzzy sets *Information and Control* **8**: 338–353.

INDEX

Note: Page references followed by '**f**' represents a figure, '**n**' represents a footnote and '**t**' represents a table.